How Computers Work

Processor and Main Memory

By

Roger Young

ISBN: 1-4033-2582-0 (Paperback)
ISBN: 1-4033-2581-2 (e-book)

This book is printed on acid free paper.

1stBooks - rev. 07/22/02

Introduction

Computers are the most complex machines that have ever been created. Very few people really know how they work. This book will tell you how they work and no technical knowledge is required. It explains the operation of a simple, but fully functional, computer in *complete* detail. The simple computer described consists mainly of a processor and main memory. Relays, which are explained, are used in the circuitry instead of transistors for simplicity. This book does not cover peripherals like modems, mice, disk drives, or monitors.

Did you ever wonder what a bit, a pixel, a latch, a word (of memory), a data bus, an address bus, a memory, a register, a processor, a timing diagram, a clock (of a processor), an instruction, or machine code is? Though most explanations of how computers work are a lot of analogies or require a background in electrical engineering, this book will tell you *precisely* what each of them is and how each of them works without requiring *any* previous knowledge of computers or electronics. However, this book starts out very easy and gets harder as it goes along. You must read the book starting at the fist page and not skip around because later topics depend on understanding earlier topics. How far you can get may depend on your background. A junior high school science background should be enough. There is no mathematics required other than simple addition and multiplication. This is a short book, but it must be studied carefully. This means that you will have to read some parts more than once to understand them. Get as far as you can. You will be much more knowledgeable about how computers work when you are done than when you started, even if you are not able to get through the whole text. This is a technical book though it is aimed at a non-technical audience. Though this book takes considerable effort to understand, it is *very* easy for what it explains. After you have studied this book, if you go back and read it, it will seem simple. Good Luck!

Table Of Contents

BASICS

Simple Circuit

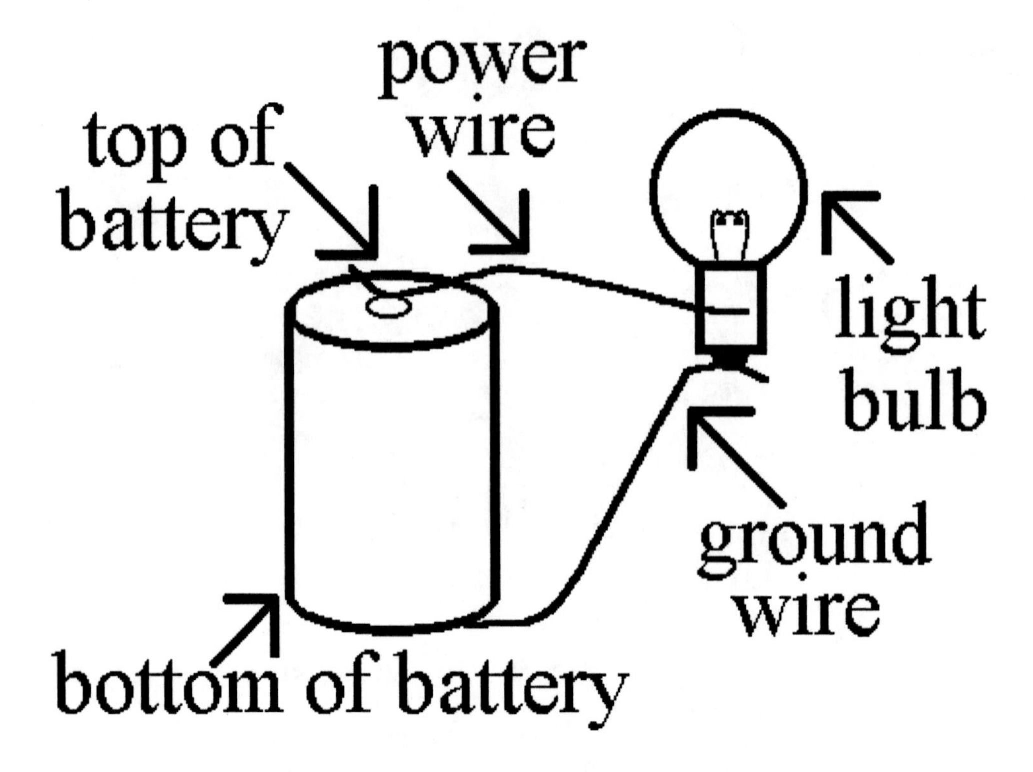

The picture above shows a 'battery' connected to a 'light bulb' by a 'power wire' and a 'ground wire.' A power wire is a wire connected directly to the top of the battery. A ground wire is a wire connected directly to the bottom of the battery. Any electrical machine is called a circuit.

Simple Diagram

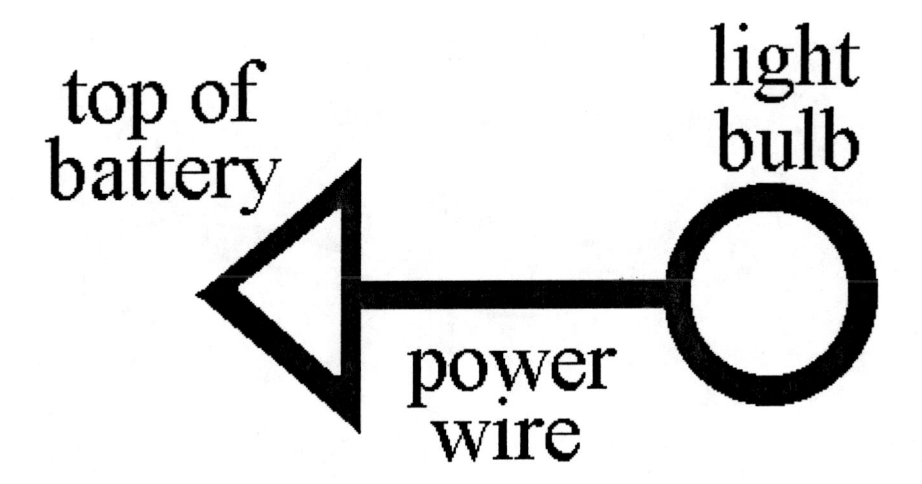

The diagram above also shows a 'battery' connected to a 'light bulb' by a 'power wire' and a 'ground wire.' This *diagram* means the same as the picture on page 2. *The ground wire is not shown* because it is assumed that one connection of every light is always connected to the bottom of the battery by a ground wire *in diagrams.* Diagrams are simpler to draw than pictures that mean the same thing.

Key Circuit

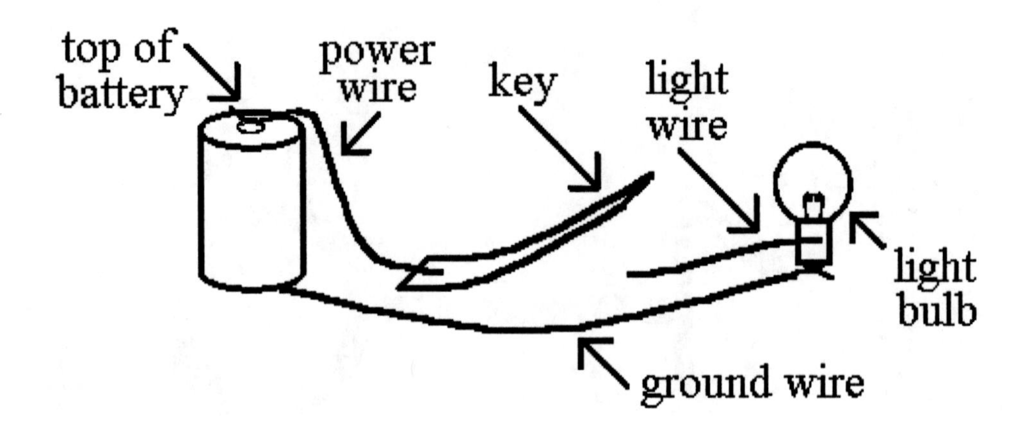

The picture above shows the 'top of' a 'battery' connected by a 'power wire' to a 'key' that is connected by a 'light wire' to a 'light bulb.'

A key is a flat piece of springy steel that is bent up so that the key only touches the wire to the key's right when the key is pressed down by someone's finger.

When someone pushes the key down, the right end of the key touches the light wire and electricity flows from the top of the battery, through the power wire, the key, and the light wire, to the light bulb, turning the light bulb on.

When the key is released, the key springs back up. Now the key does *not* touch the light wire and electricity can *not* get from the key to the light wire to the light bulb so that the light bulb goes *off*.

Key Diagram

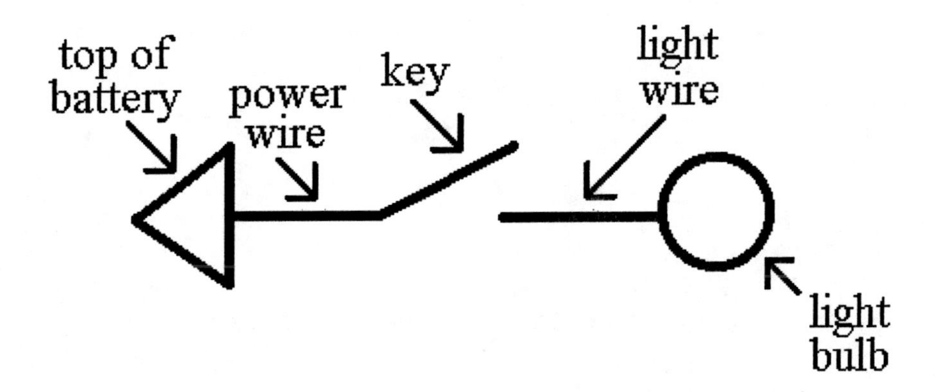

The diagram above shows the same circuit as the preceding picture.

Again, there is also a wire from the other connection of the light bulb back to the bottom of the battery, but that wire does not need to be shown because the other connection of every light is connected to the bottom of the battery and you know the ground wire is there without drawing it.

Electromagnet

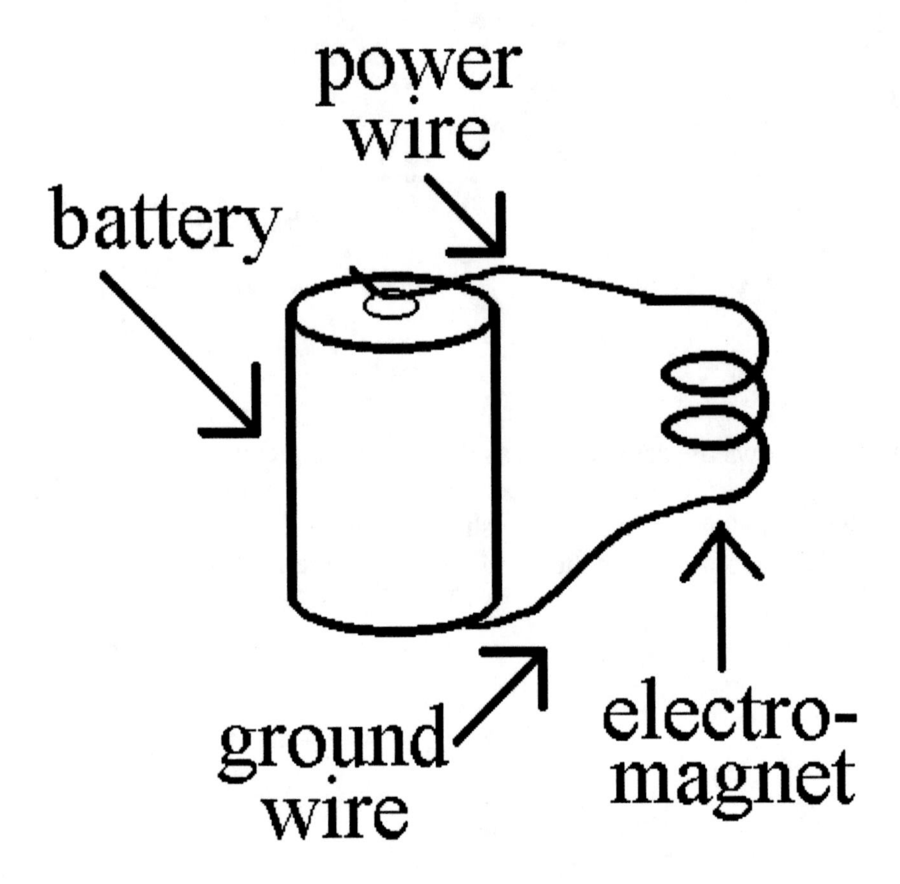

The picture above shows the top of a battery connected by a wire to an electromagnet.

An electromagnet is a coil of (plastic coated) wire. An electromagnet becomes magnetic when electricity goes through it, just as a light bulb glows when electricity goes through the light bulb.

The wire that makes up the coil of wire that is the electromagnet has two ends (connections). There is also a 'ground wire' from the other connection of the electromagnet back to the bottom of the battery.

Electromagnet Diagram

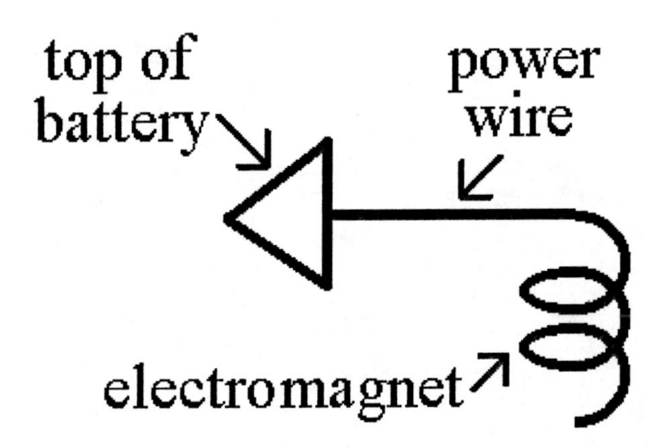

The diagram above shows the same circuit as the preceding picture.

The wire that makes up the coil of wire that is the electromagnet has two ends (connections). There is also a ground wire from the other connection of the electromagnet back to the bottom of the battery, as in the picture, but that wire does *not* need to be shown because the other connection of *every* electromagnet is connected to the bottom of the battery.

Relay

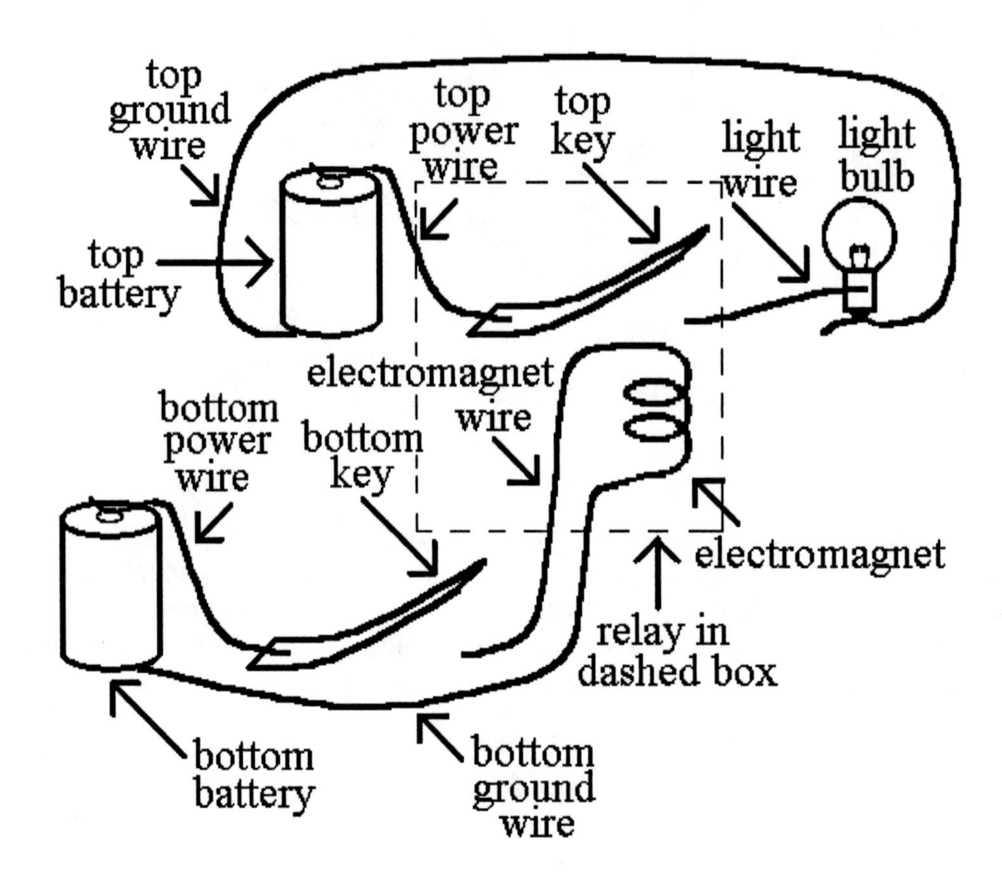

The picture above shows a 'bottom key' that controls an electromagnet.

The electromagnet, in turn, controls the top key. A key and the electromagnet that controls it are, *together*, called a *relay*. The relay is in the dashed box.

When the bottom key is pressed, the electromagnet is powered and the electromagnet becomes magnetic. That makes the electromagnet attract the top key and pull the top key down just like a finger can push a key down. A magnet (or a powered electromagnet) attracts the top key because the top key is made of steel. A magnet (or a powered electromagnet) does not attract the wires because the wires are made of copper.

Important: The electromagnet does *not ever* touch the top key. *No* electricity can go from the electromagnet to the wires attached to the top key.

A computer is almost entirely made up of a lot of relays (today, transistors) connected by wires. Just how the relays are connected and just what they do is the main subject of this book. Other concepts, especially programming, will also be explained.

(Today, transistors are used instead of relays for lower cost and greater speed. The design remains practically the same, however. Relays are easier to understand and, so, will be used in this explanation.)

Relay Diagram

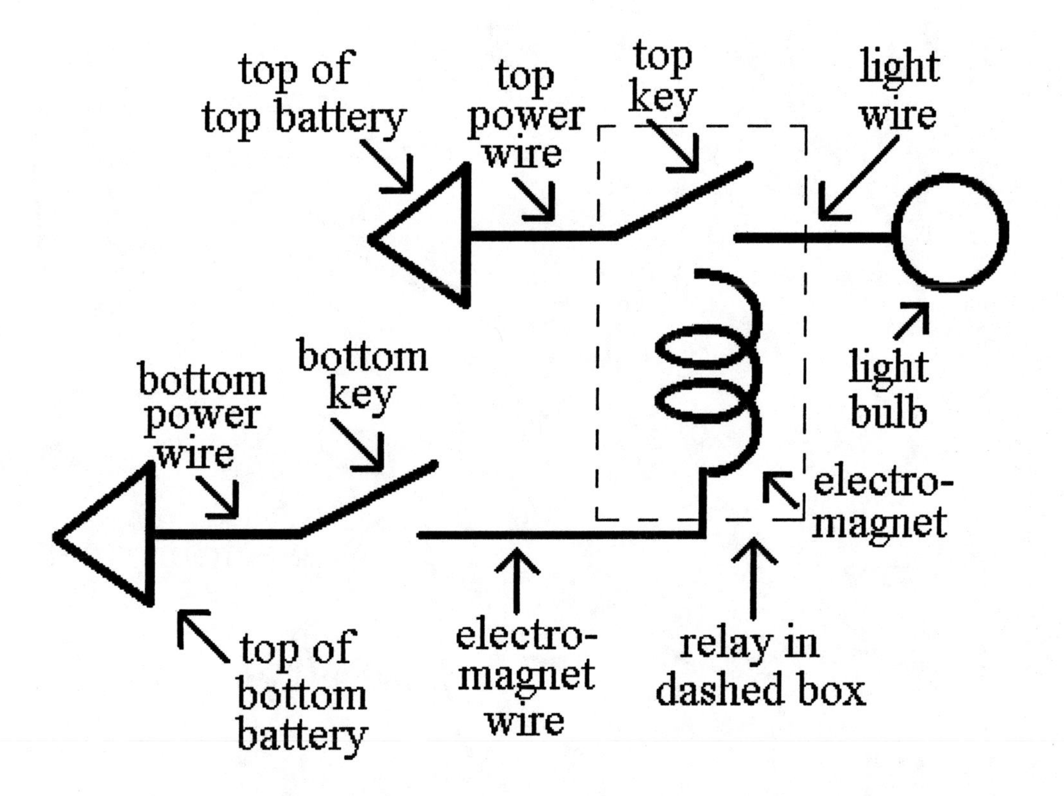

The diagram above shows the same circuit as the previous picture in a different way.

One Battery And Touching Wires

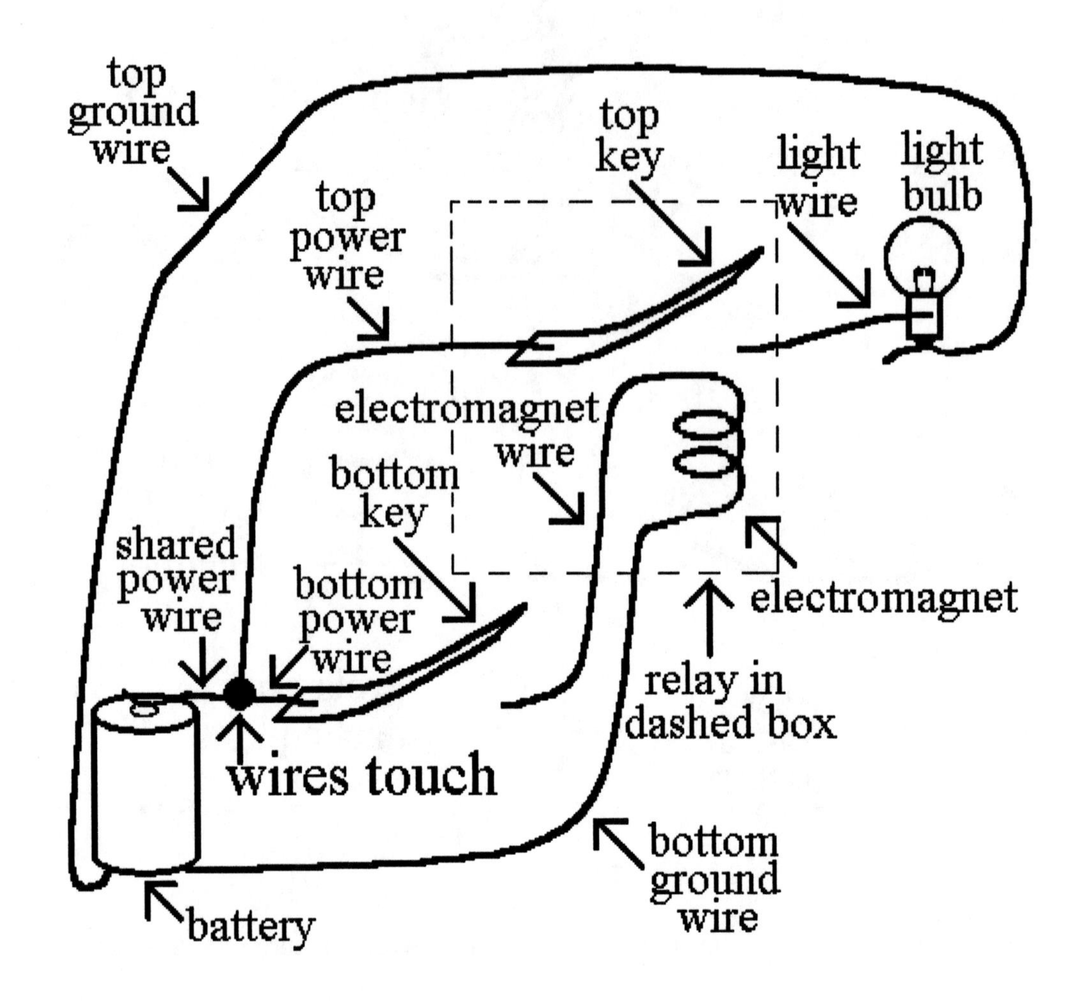

In this picture, only *one battery* powers all the circuitry in the previous picture. Note the symbol for wires that touch.

One Battery and Connected Wires Diagram

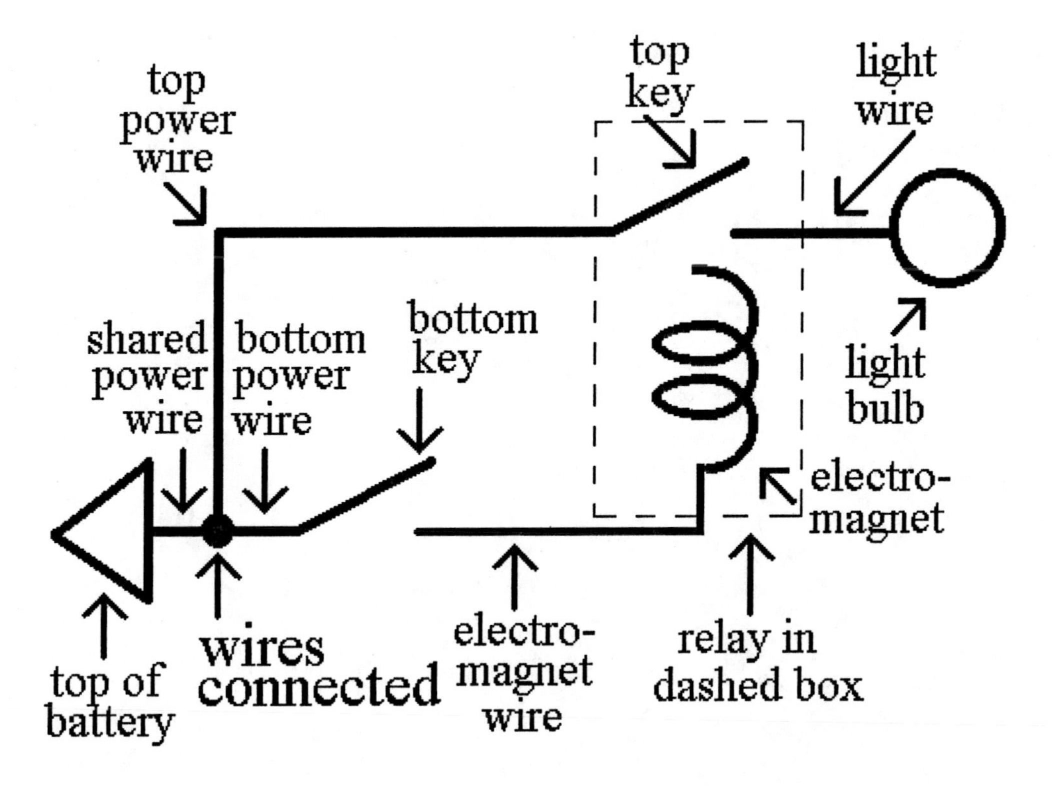

This diagram shows the same circuit as the previous picture in a different way. Touching wires are connected wires.

Loop

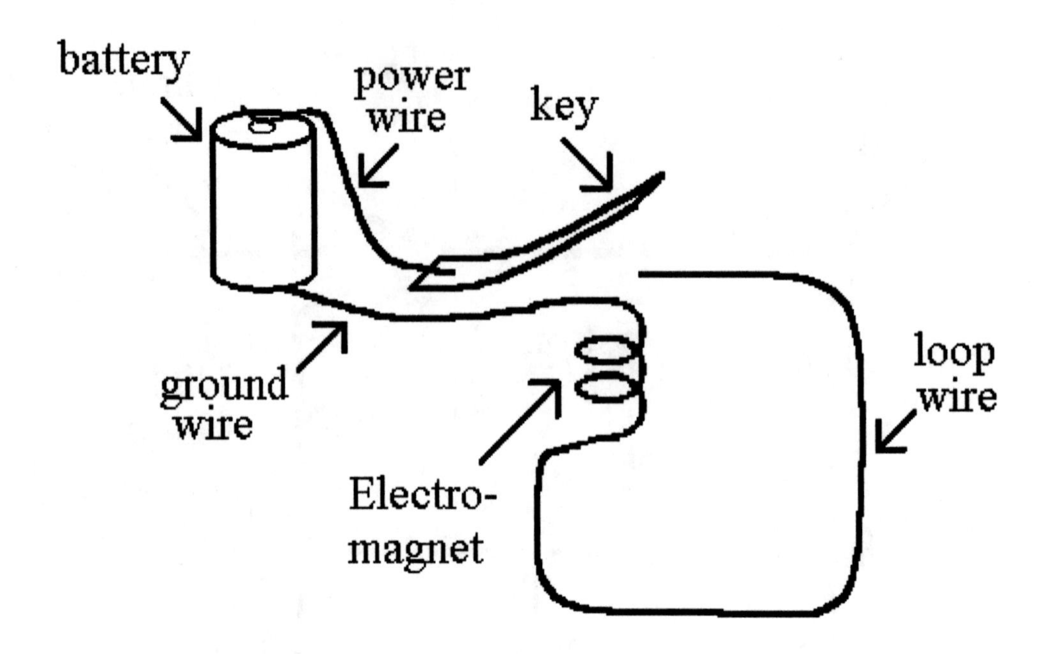
battery
power wire
key
ground wire
Electro-magnet
loop wire

Loop Diagram

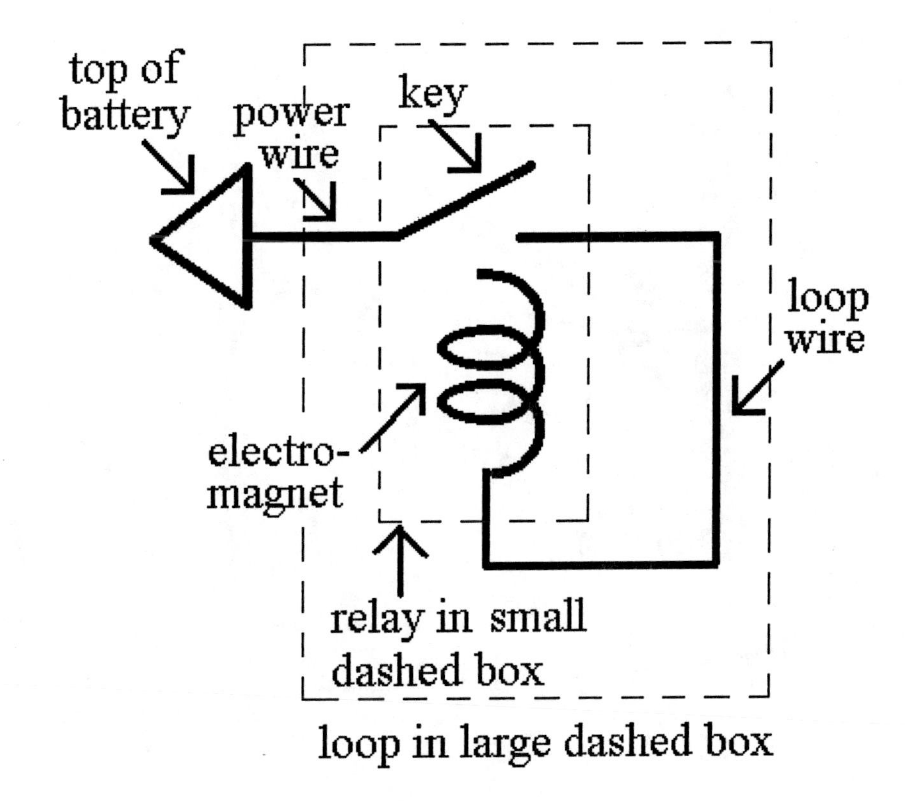

The picture and diagram at left show a relay that controls its *own* electromagnet! The square of wire that takes electricity from the key of the relay to the electromagnet of the same relay is called a 'loop.'

No electricity can get from the top of the battery to the electromagnet because the key is up. However, if someone presses the key, then electricity *can* get to the electromagnet. Then, the *electromagnet* will hold the key down - *even if the person lets go of the key!* So we say that the loop *remembers* that the key was pressed. Remember that the key normally springs up because it is springy and bent upward.

Similarly, if someone then lifts up the key (A person is much stronger than a little electromagnet.), then no electricity will reach the electromagnet and the key will remain up even after the person releases the key. So we say that the loop *remembers* that the key was lifted up.

Most relays in a computer are used to make loops, or connect the loops together.

Pixel

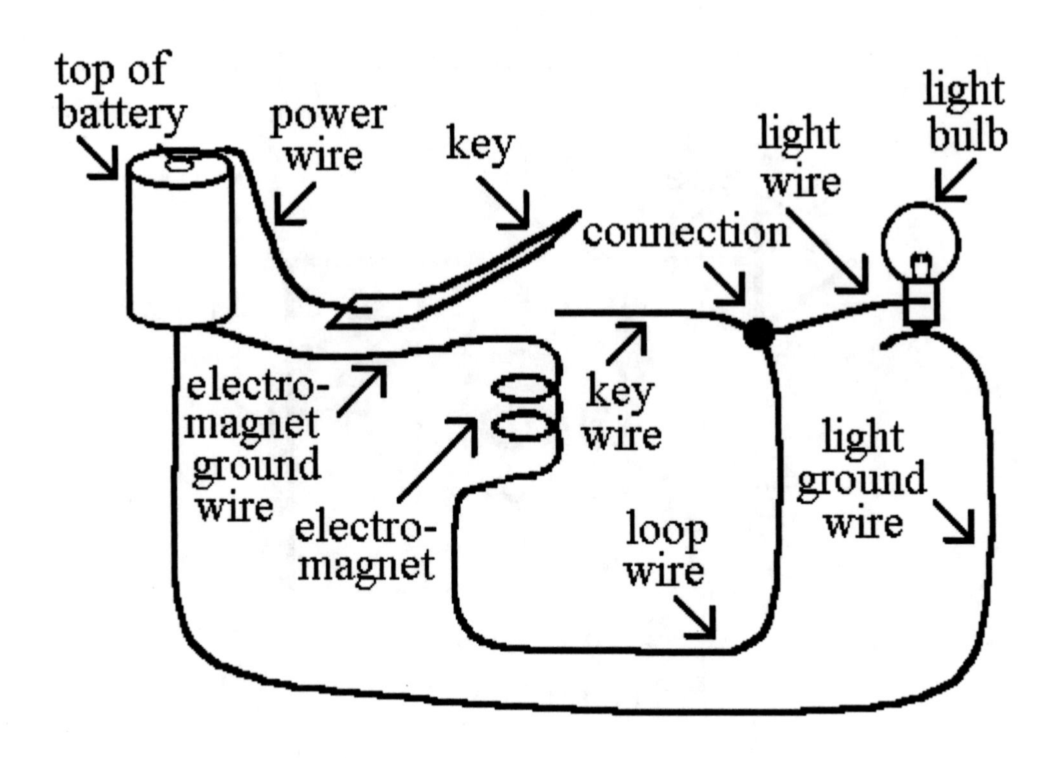

top of battery
power wire
key
light wire
light bulb
connection
electro-magnet ground wire
key wire
electro-magnet
light ground wire
loop wire

Pixel Diagram

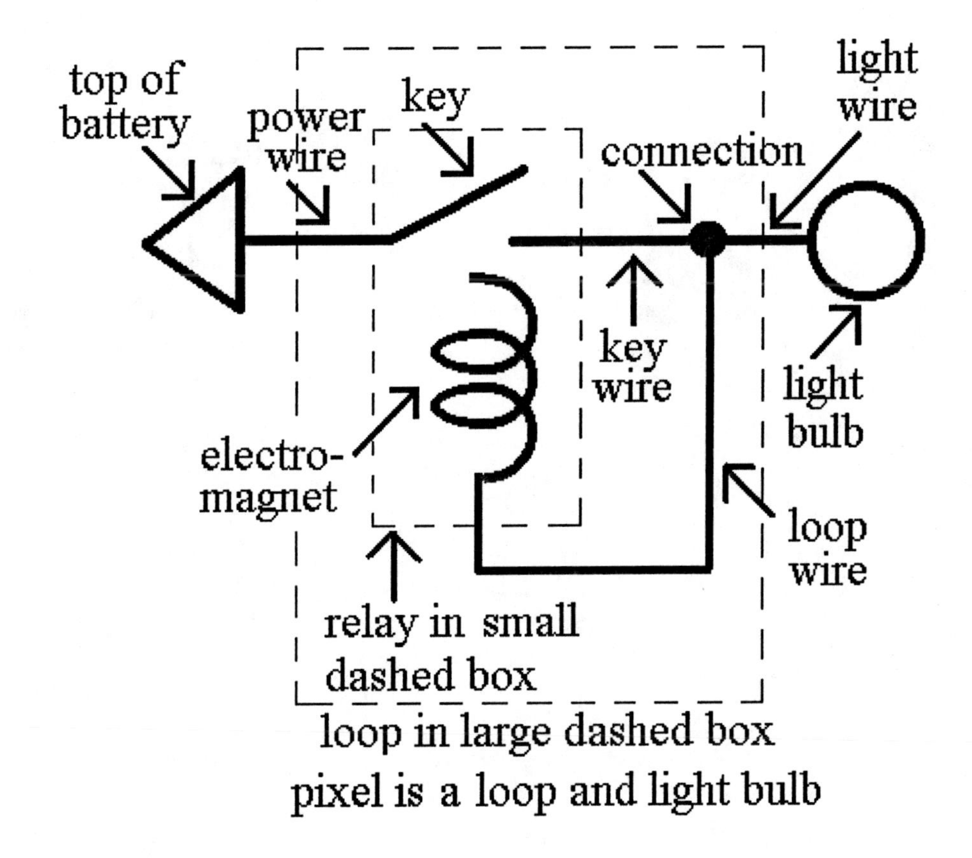

The picture and diagram above show a loop that controls a light bulb. A light bulb that is controlled by a loop is called a 'pixel.'

In a diagram, where a horizontal wire and a vertical wire meet, *without crossing*, there *is* a connection of the two wires.

Therefore, when the key is pressed, electricity can flow from the top of the battery, through the key, to both the light and the electromagnet. When the key is down and the light bulb is glowing, one says that the loop has value '1' and the pixel is 'on.' The loop has value '1' even if there is not a light bulb, just so the loop wire has electricity going through it, to the electromagnet, because the key is down.

When the key is *up* and the light bulb is *not* glowing, one says that the loop has value '0' and the pixel is 'off.' The loop has value '0' even if there is not a light bulb - just so the loop wire does *not* have electricity going through it (because the key is up).

Normally Closed Key

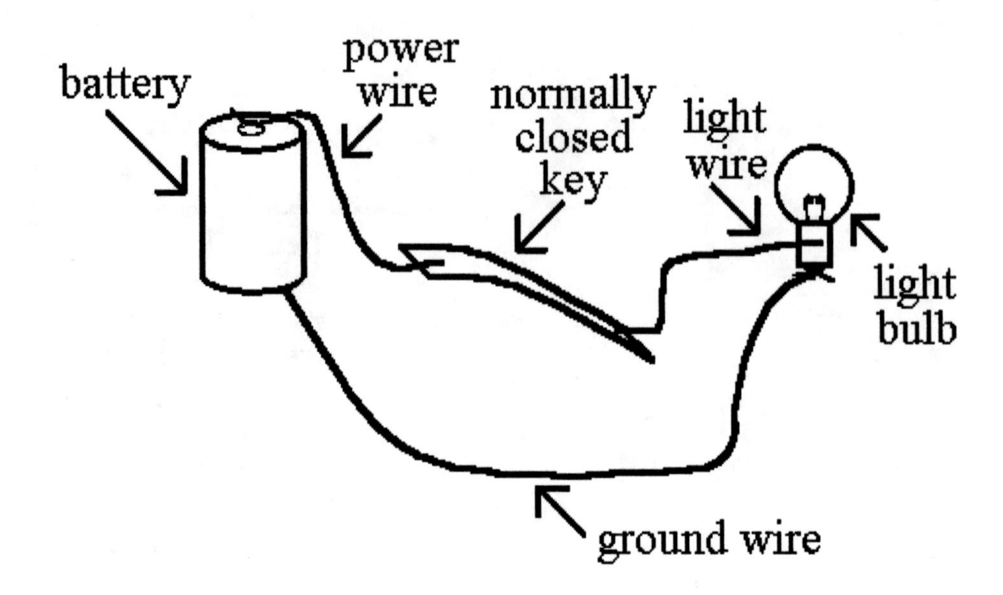

Normally Closed Key Diagram

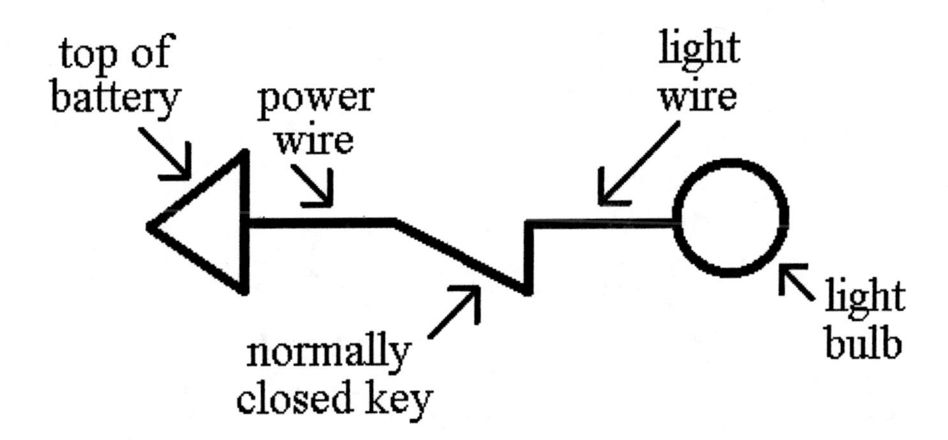

The picture and diagram at left show the top of a battery connected by a wire to a normally closed key, that is connected by another wire to a light bulb.

A diagram of an electrical machine is called a circuit diagram, a diagram, a schematic (pronounced ske-ma'-tic) diagram, or just a schematic.

The normally closed key is different from the keys described previously. The normally closed key is also a springy piece of steel, but is bent so that it normally *is* connected to the right wire. Therefore, the light bulb in the circuit above is normally on. However, if you push *down* on the normally closed key, the light bulb becomes disconnected from the '*power* wire' and the light goes out.

A key is called 'closed' when electricity can flow through it from a wire on its left to a wire on its right.

A key is called 'open' when electricity *can't* flow through it from a wire on the left to a wire on the right.

A normally closed key is normally closed, but is open when you push it down.

A normally open key is normally open, but is closed when you push it down.

A relay is called closed if its key is closed.

A relay is called open if its key is open.

An electromagnet is called 'powered' if the electromagnet is connected to the top of a battery, *even* if that electromagnet is connected to the top of the battery through a series of *closed* keys. In fact, any piece of wire is called 'powered' if that piece of wire is connected to the top of a battery, *even* if that piece of wire is connected to the top of the battery through a series of *closed* keys.

Any piece of wire that is powered is said to have value '1.'

Any piece of wire that is *not* powered is said to have value '0.'

The values of the wire in a loop as described previously are a special case of these rules for assigning values to wires.

Normally Closed Relay

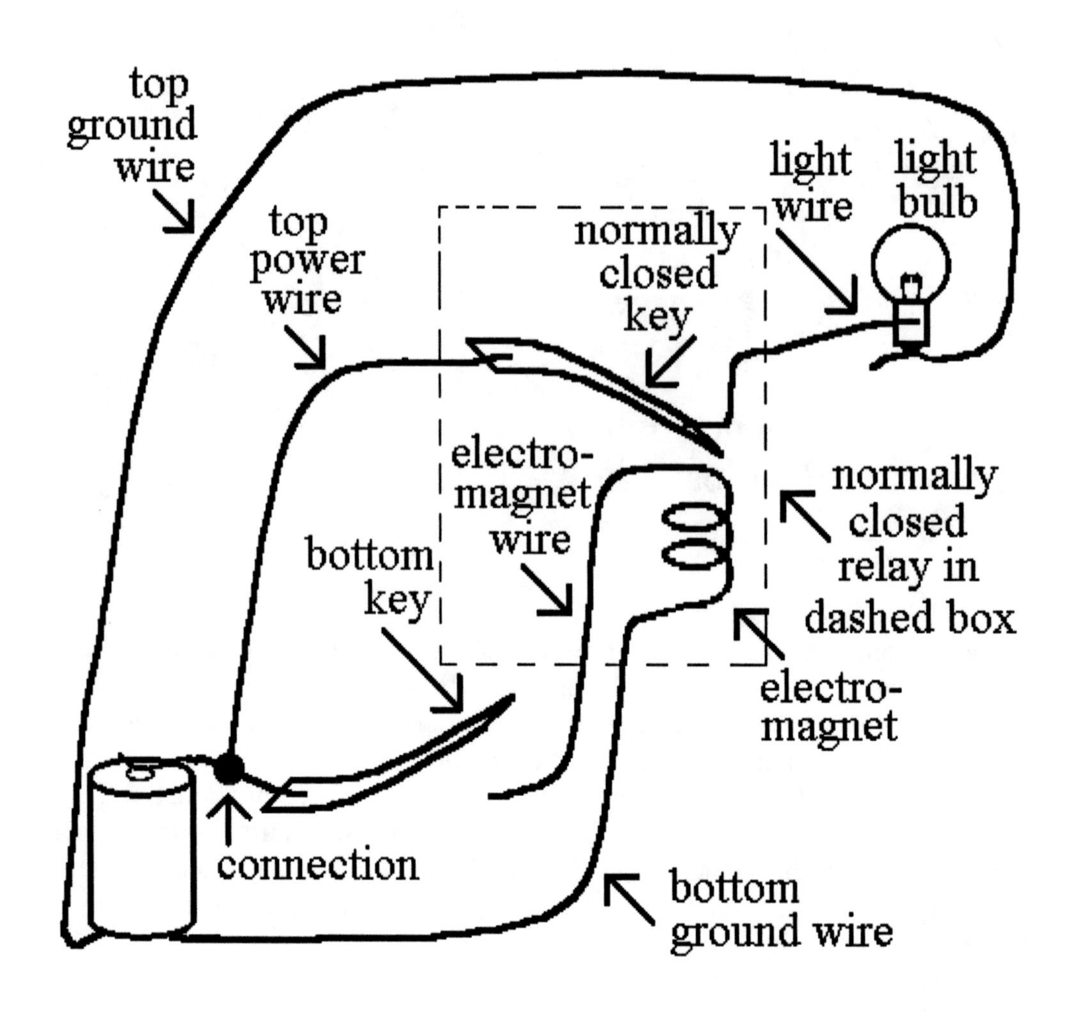

Normally Closed Relay Diagram

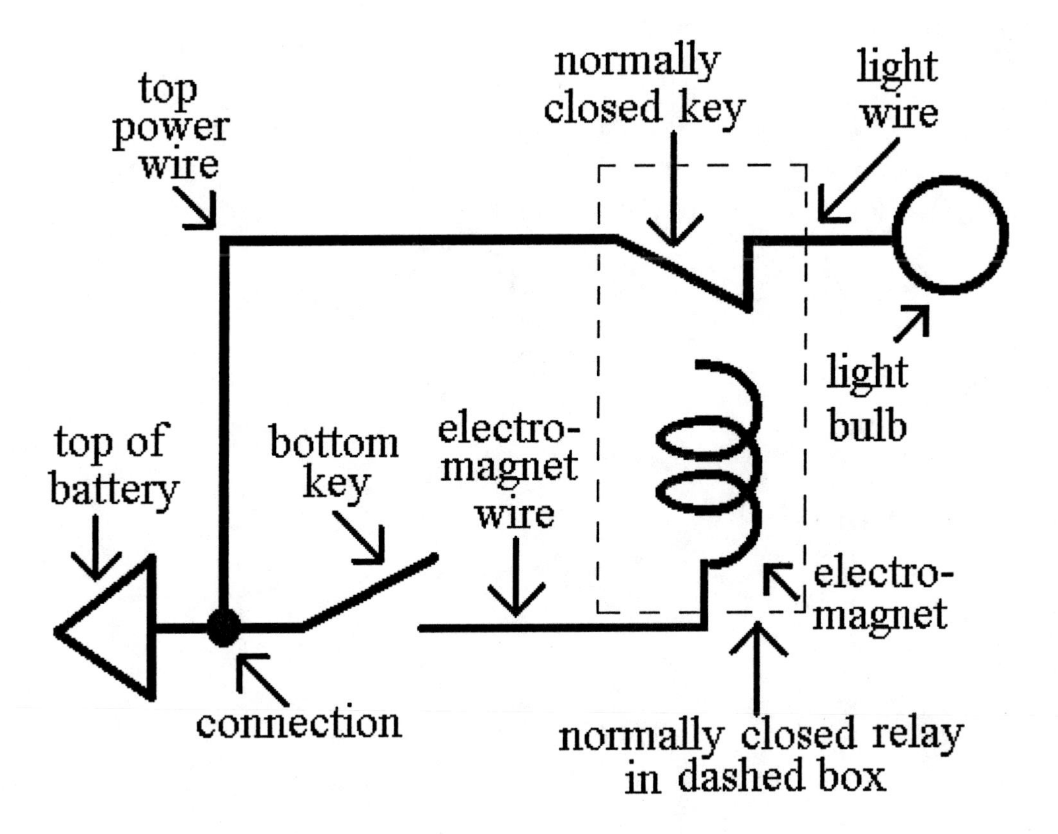

The preceding picture and diagram show a bottom key that controls an electromagnet. The *electromagnet*, in turn, controls the *top, normally closed key*. A normally closed key and the electromagnet that controls it are, *together*, called a normally closed relay.

When the bottom key is pressed, the electromagnet is powered and the electromagnet becomes magnetic. That makes the electromagnet attract the top, normally closed key and pull the top, normally closed key down, just like a finger can push a normally closed key down. A magnet (or a powered electromagnet) attracts the normally closed key because the normally closed key is made of steel. When the bottom key is pressed, the light turns *off*.

In other words, when the bottom key is pressed, the electromagnet energizes, disconnecting the top key.

Clear Key

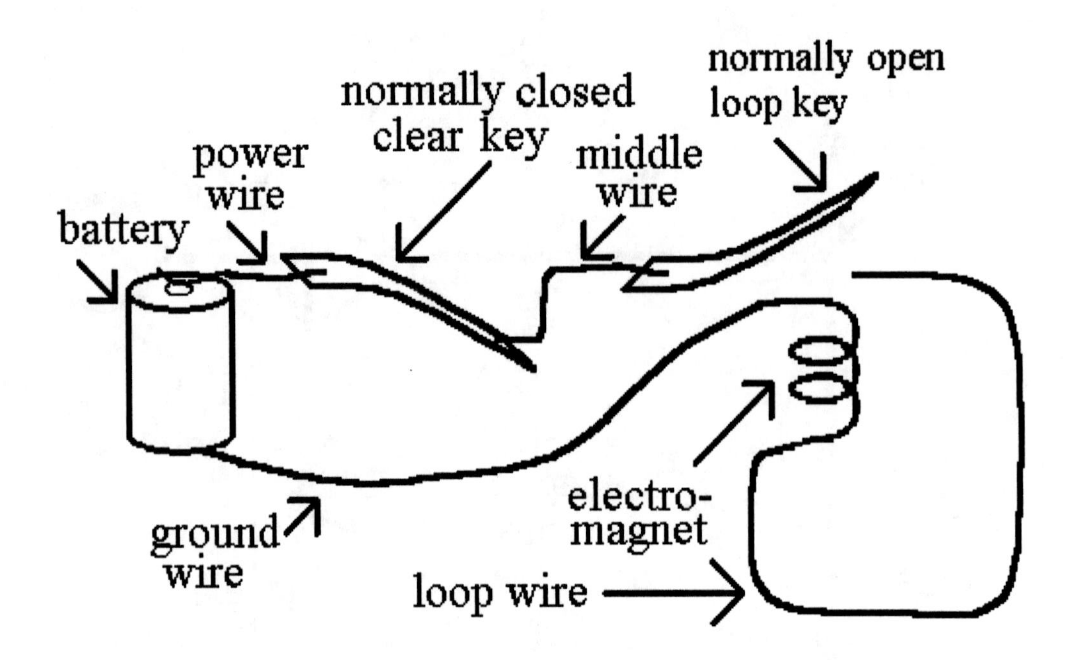

normally open
loop key
normally closed
clear key
power
wire
middle
wire
battery
electro-
magnet
ground
wire
loop wire

Clear Key Diagram

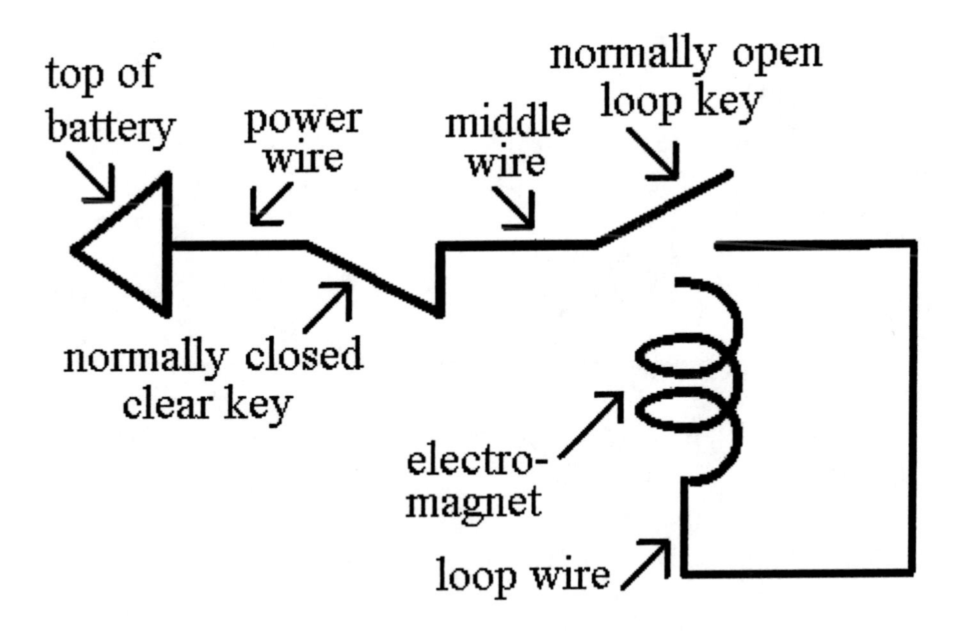

The picture and diagram above show a loop as before, but a normally closed key has been added. As long as the normally closed key is closed, the loop works as before.

However, if the normally closed key is pressed, then the normally closed key will be open and electricity will not reach the electromagnet, so the electromagnet will not be magnetic, and the normally open key will pop up if it was down. If the normally open key already was up, it will stay up.

Therefore, pressing the normally closed key will clear the value of the loop to '0.' Therefore, this normally closed key is called the 'clear key' for the loop.

Loop to Loop Data Transfer

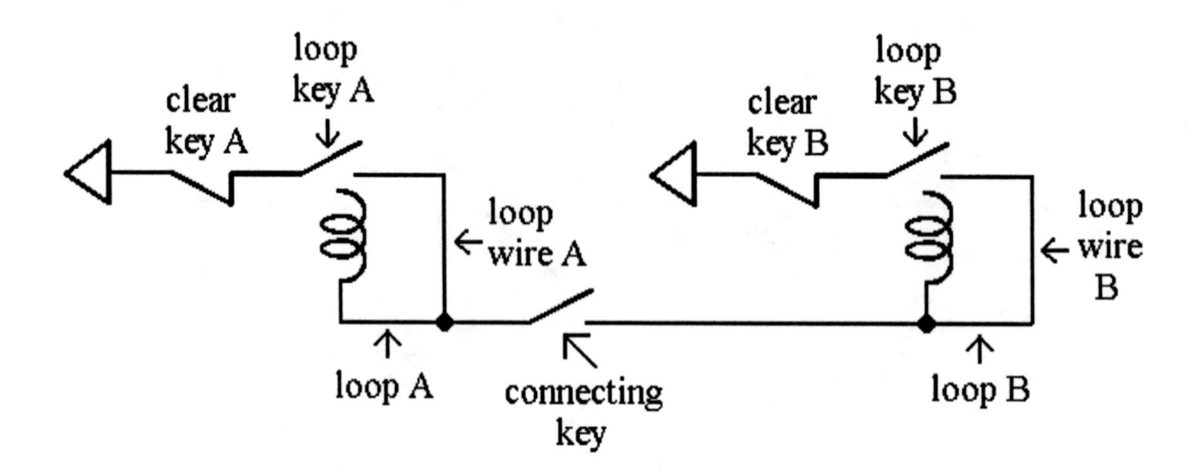

In the circuit above, the 'connecting key' connects loop A and loop B. Both loops have value 0. Temporarily pressing 'loop key A' gives the value 1 to loop A. Now, temporarily pressing the 'connecting key' will make loop B have value 1. That is because when loop A has value 1, loop key A is closed, loop wire A has value 1, and when the connecting key is closed, electricity can reach the electromagnet of loop B, giving loop B value 1.

However, if loop A has value 0, and loop B has value 0, and the connecting key is pressed, then both loops keep their values of 0.

Therefore, *if one presses 'clear key B' to clear loop B to value 0, and then temporarily presses the connecting key, whatever value is in loop A will be copied to loop B.* Then loop A and loop B will have the same value.

Oscillator

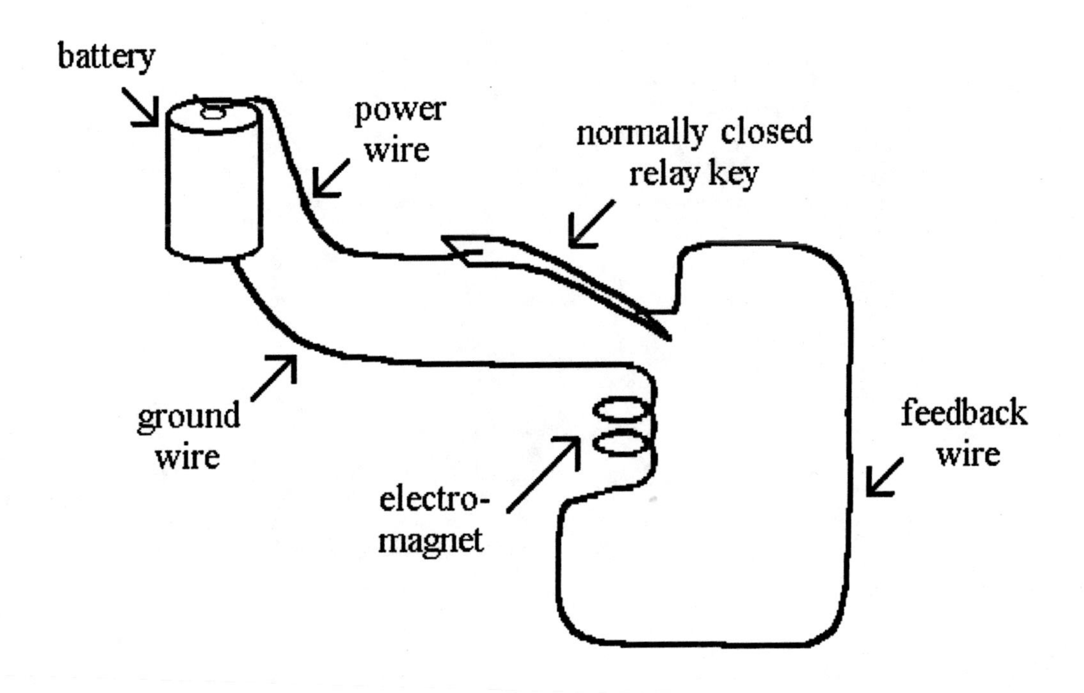
battery
power
wire
normally closed
relay key
ground
wire
feedback
wire
electro-
magnet

Oscillator Diagram

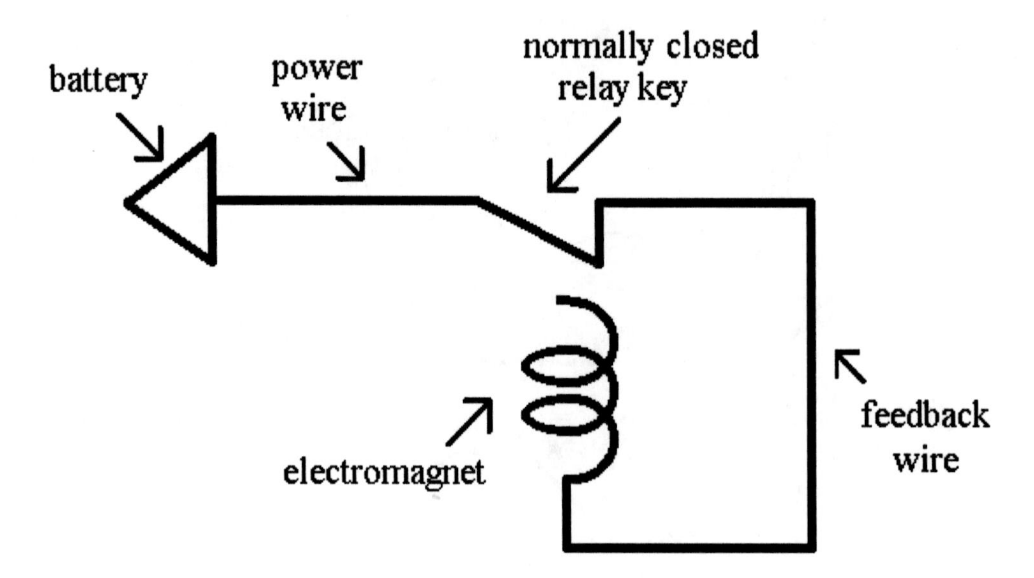

The picture and diagram at left show a normally closed relay that controls its *own* electromagnet. The square of wire that takes electricity from the normally closed key of the relay to the electromagnet of the same normally closed relay is called a feedback wire. (Notice that this circuit is different from a loop circuit, which uses a normally open relay.) This circuit is called an oscillator because the relay oscillates (changes back and forth) between open and closed.

Electricity can get from the top of the battery, through the *closed*, normally closed relay key to the electromagnet. The electromagnet then pulls the normally closed key down and *opens* the normally closed key. Because the normally closed key is now open, no electricity can get to the electromagnet. The electromagnet now no longer attracts the normally closed key and the normally closed key closes.

Thus, the normally closed key repeatedly opens and closes without anyone touching the key. The feedback wire gets value 1, then value 0, then value 1, etc. It takes a relay about a hundredth of a second to change values.

Just as a normal loop is the basis of a computer memory, this feedback circuit is a key part of a computer's clock. A computer's clock is a circuit that repeatedly generates signals (1 and 0 values).

Keys in Series

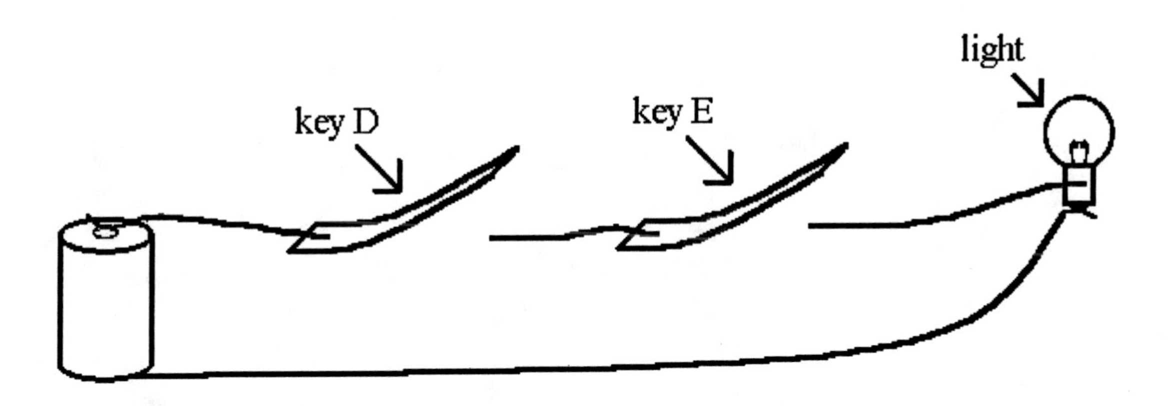

Keys in Series Diagram

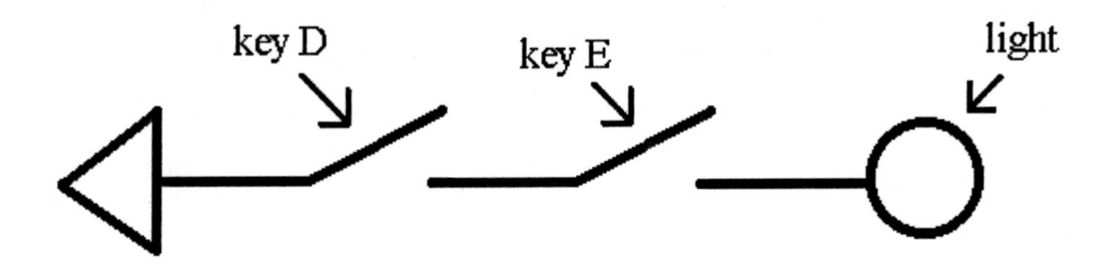

In the picture and diagram above, one must press both 'key D' AND 'key E' to turn the light on.

AND Gate Circuit

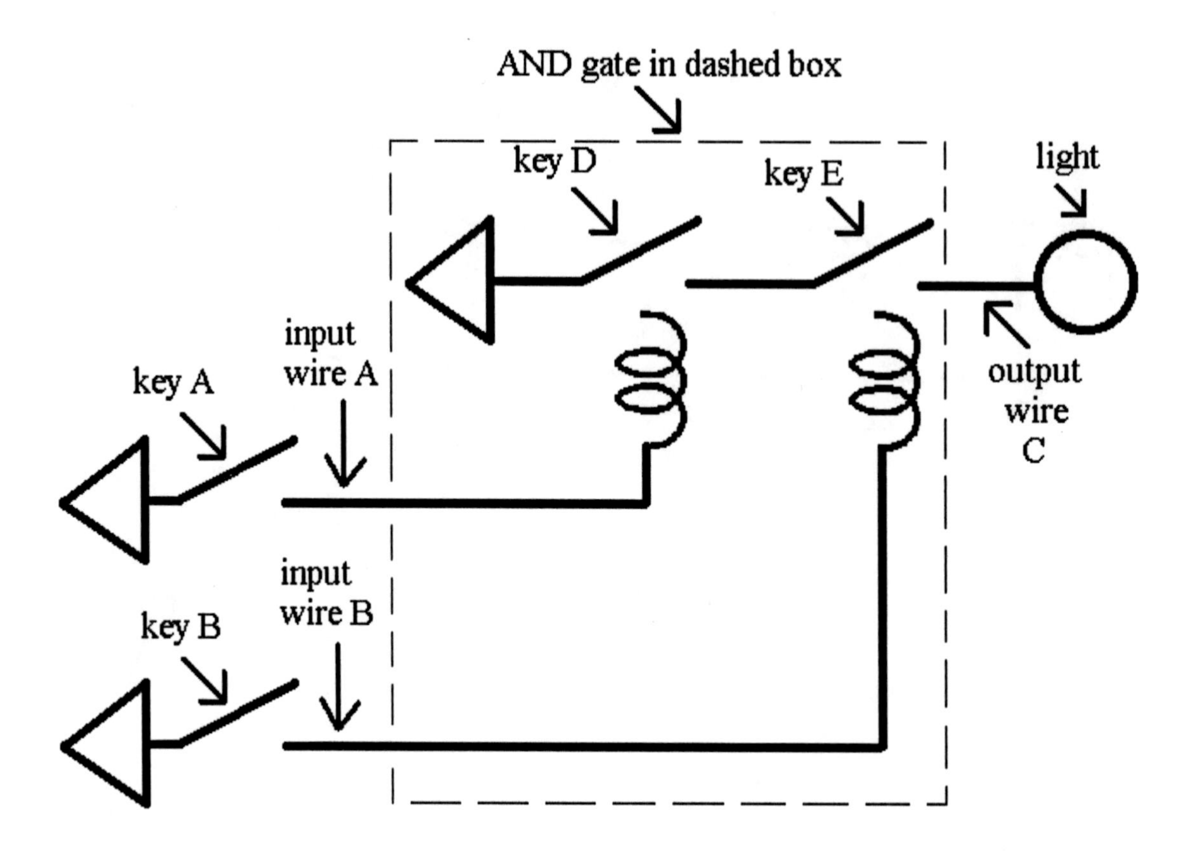

In the circuit above, the three triangles are all the top of the *same* battery. When 'key D' AND 'key E' close, then the light comes on. When 'key A' is pressed, then 'key D' closes. When 'key B' is pressed, then 'key E' closes. Therefore, when 'key A' and 'key B' are pressed, the light turns on. Another way of describing the operation of the circuit is to say that 'output wire C' gets value 1 only when 'input wire A' gets value 1 AND 'input wire B' gets value 1.

The following table also shows that 'output wire C' has value 1 only when both 'input wire A' has value 1 AND 'input wire B' has value 1.

AND gate truth table		
A	B	C
0	0	0
0	1	0
1	0	0
1	1	1

AND Gate Circuit with Symbol

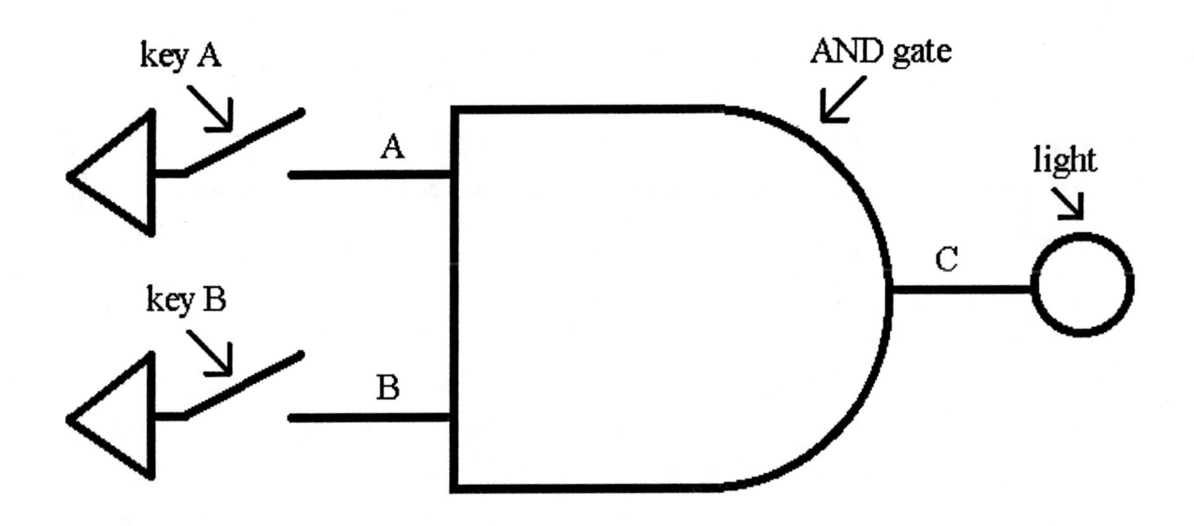

The diagram above shows a circuit with the symbol for an 'AND gate' which is shown, alone, below.

AND Gate Symbol

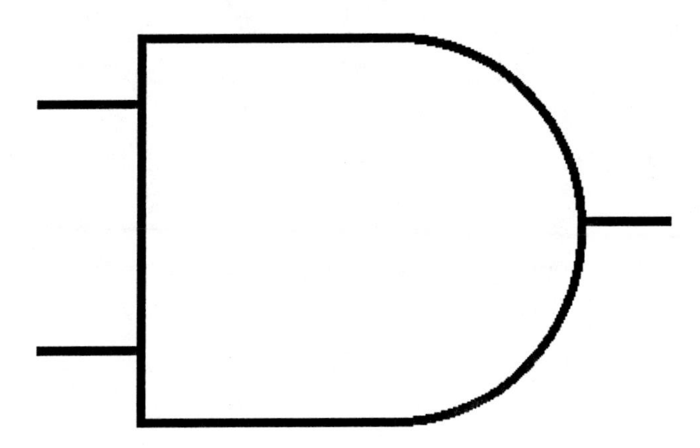

The light in the circuit below only comes on whey key D, key E, AND key F are *all* pressed.

Three Keys in Series

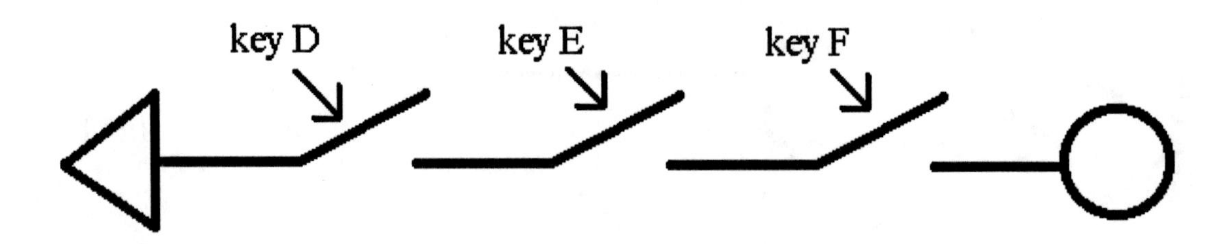

Keys in Parallel

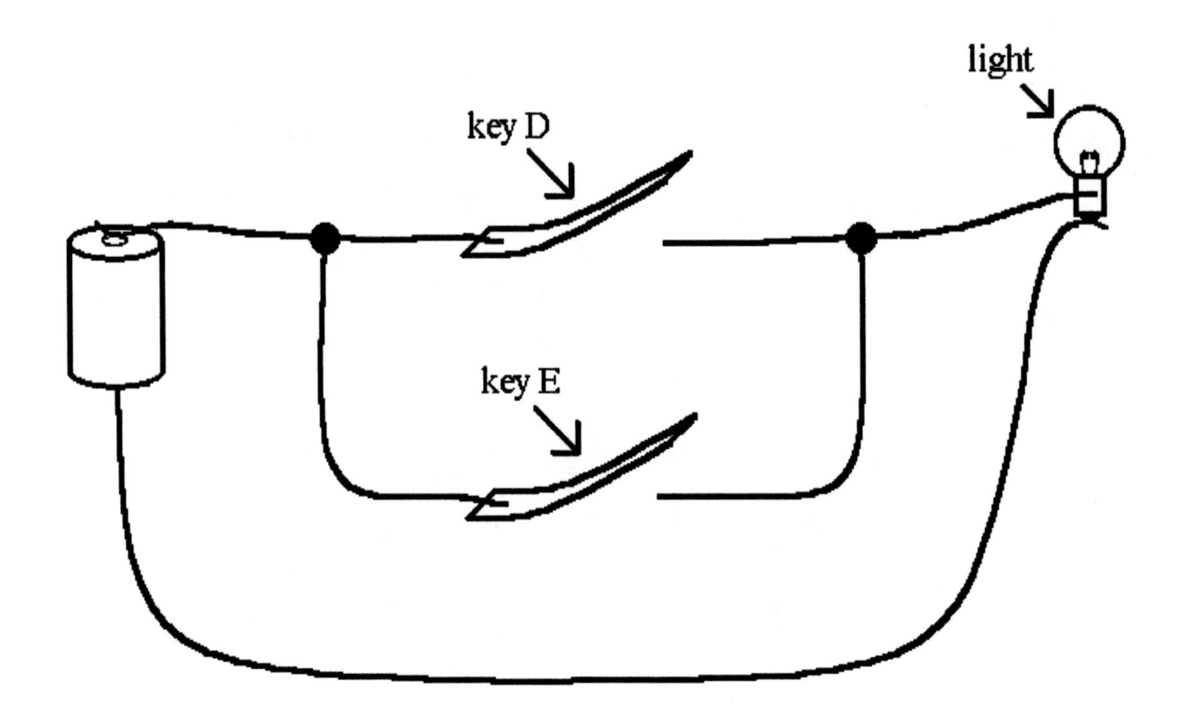

Keys in Parallel Diagram

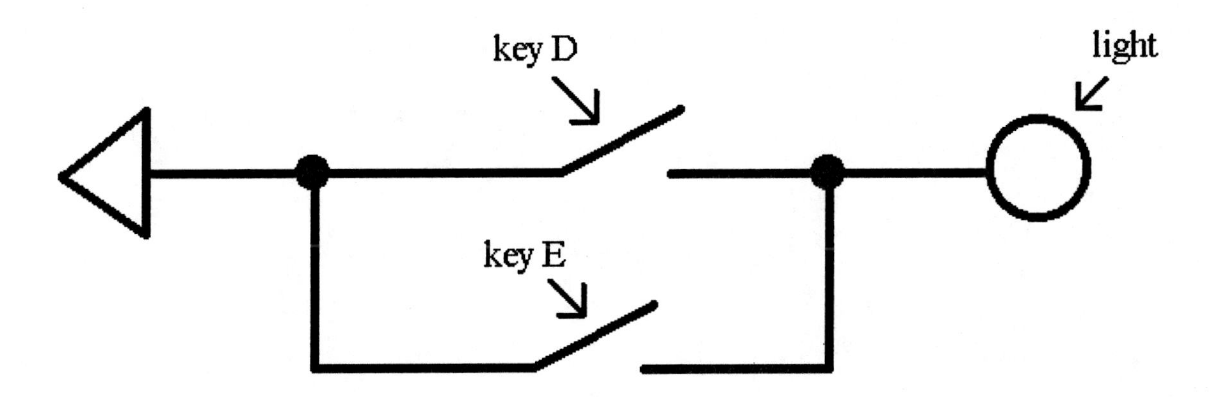

In the picture and diagram above, one need only press *either* ‘key D’ OR ‘key E’ (or both) to turn the light on.

OR Gate Circuit

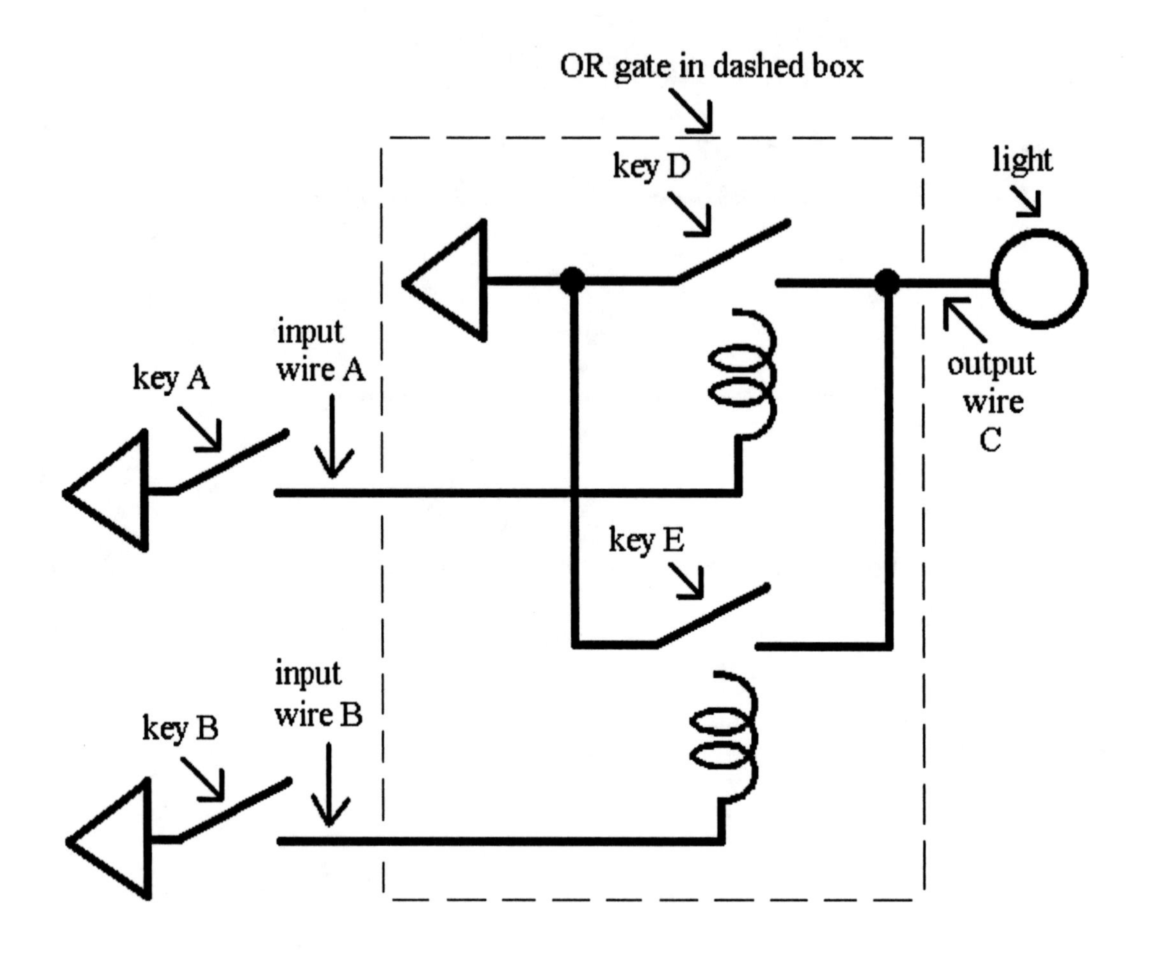

In the circuit above, when 'key D' OR 'key E' (or both) closes, the light comes on. When 'key A' is pressed, then 'key D' closes. When 'key B' is pressed, then 'key E' closes. Therefore, when 'key A' OR 'key B' is pressed, the light turns on. Another way of describing the operation of this circuit is to say that 'output wire C' gets value 1 only when 'input wire A' has value 1 OR 'input wire B' has value 1.

The following table also shows that 'output wire C' gets value 1 only when either 'input value A' has value 1 OR 'input wire B' has value 1.

OR gate truth table		
A	B	C
0	0	0
0	1	1
1	0	1
1	1	1

OR Gate Circuit with Symbol

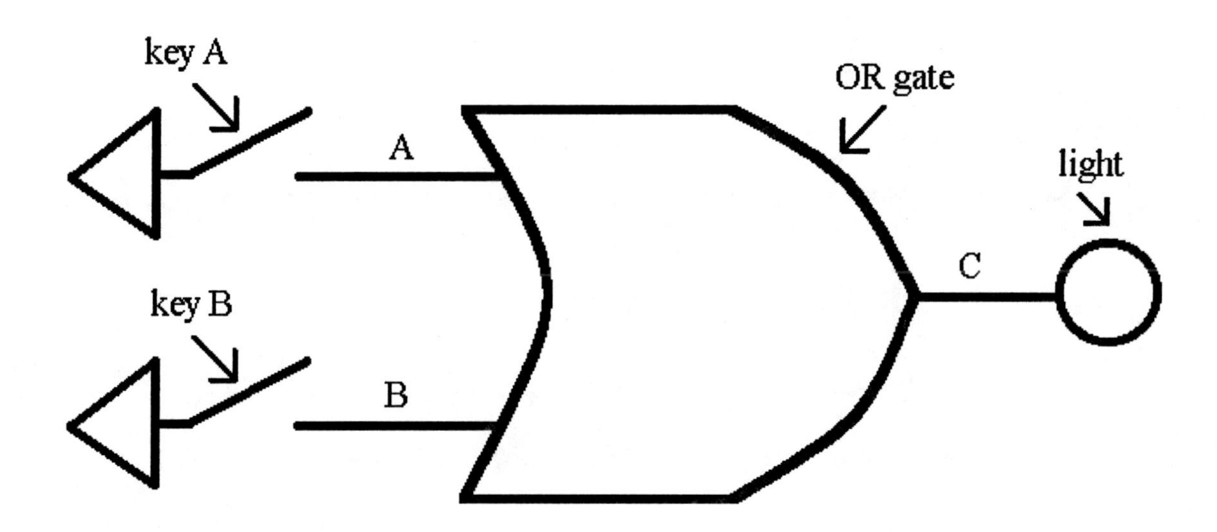

The diagram above shows a circuit with the symbol for an 'OR gate' which is shown alone, below.

OR Gate Symbol

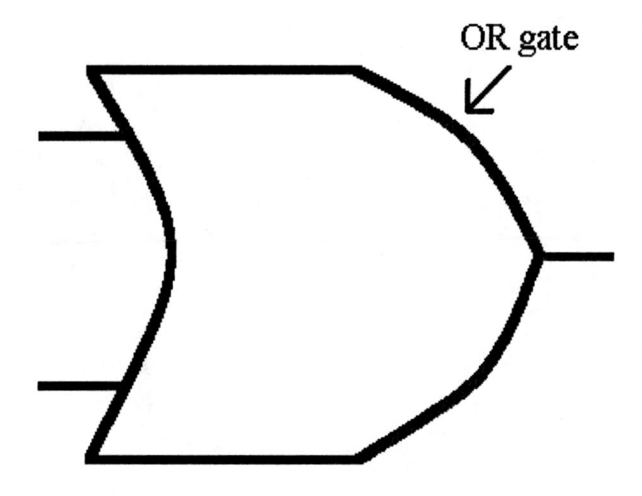

Three Keys in Parallel

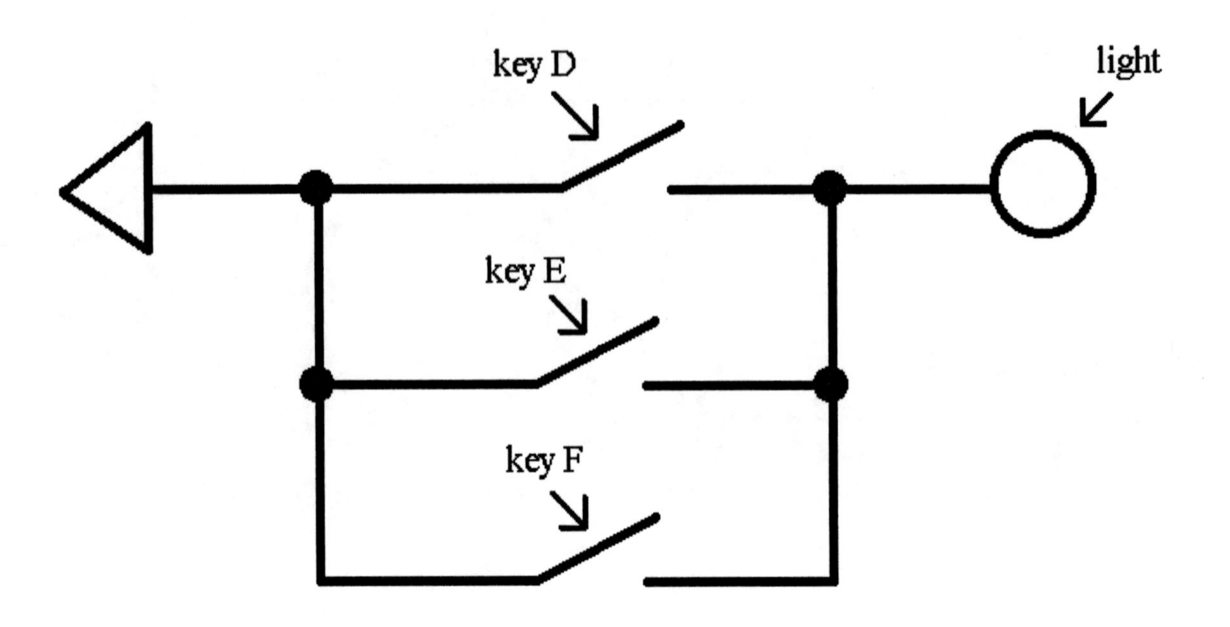

The light in the circuit above turns on when key D, key E, OR key F is pressed.

Normally Closed Key

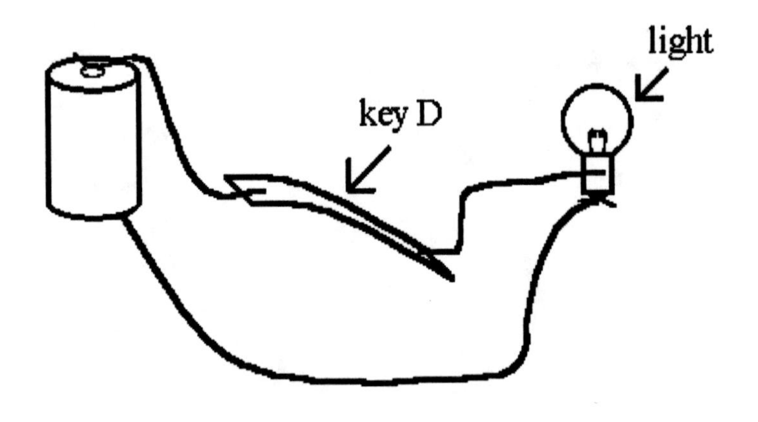

Normally Closed Key Diagram

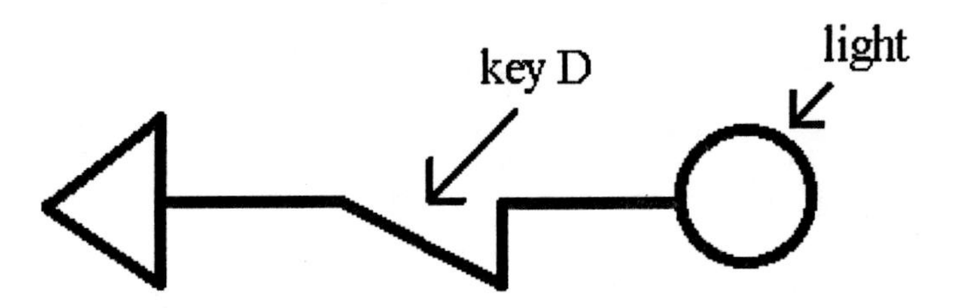

In the picture and diagram above, the light is on, as we have seen before. One must press the normally closed key D down to turn the light *off*.

NOT Gate Circuit

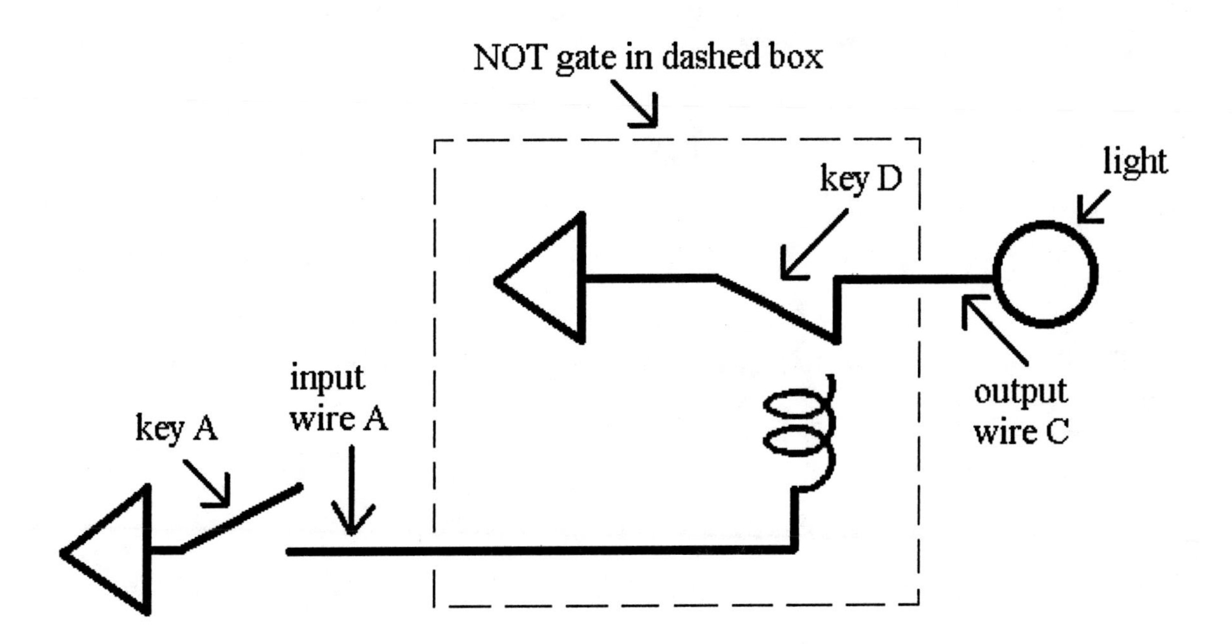

In the circuit above, the triangles are both the top of the *same* battery. When 'key A' is pressed, 'key D' is pulled down and the light goes *off*. That is, when 'key A' is pressed, normally closed 'key D' opens. Therefore, when 'key A' is pressed, the light goes *off*. Another way of describing the operation of the circuit is to say that 'output wire C' gets value 0 when 'input wire A' gets value 1. 'Output wire C' gets value 1 when 'input wire A' gets value 0.

The following table also shows that 'output wire C' gets value 0 only when 'input wire A' gets value 1.

NOT gate truth table	
A	C
0	1
1	0

NOT Gate Circuit with Symbol

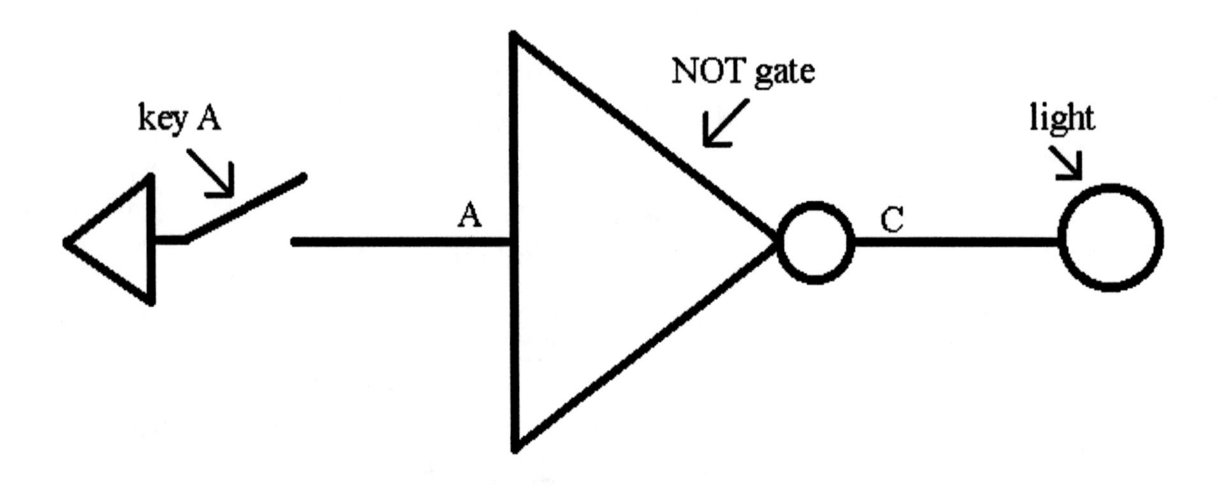

The diagram above shows a circuit with the symbol for a 'NOT gate' which is shown alone, below.

NOT Gate Symbol

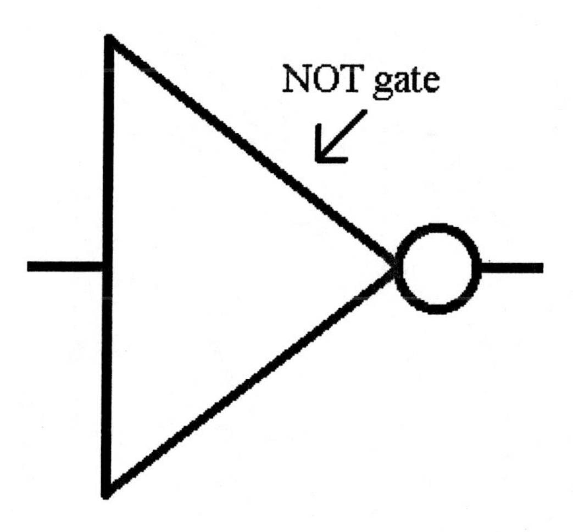

Interconnected Gates

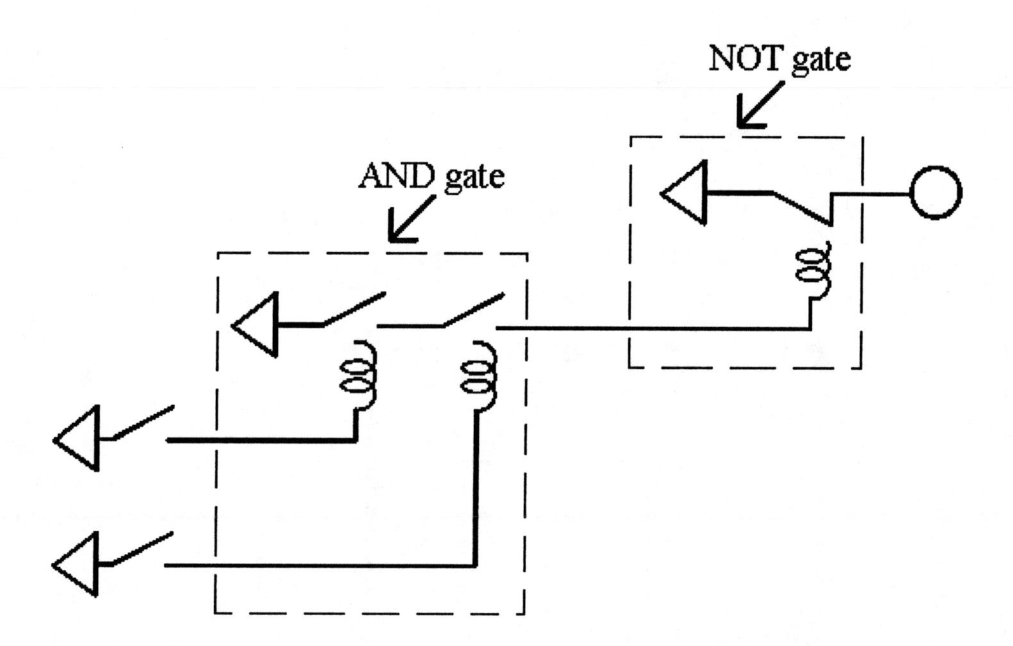

The diagram above shows that the output of an AND gate can be the input for a NOT gate. The circuit above can also be represented with gate symbols as below.

Interconnected Gates with Symbols

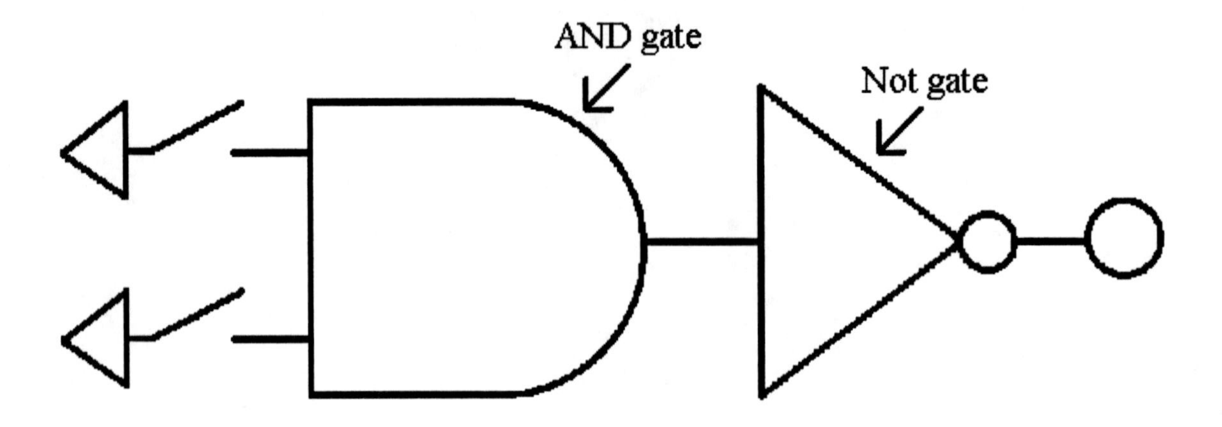

A 'NAND gate' can be constructed from an AND gate followed by a NOT gate as indicated below.

Constructed NAND Gate

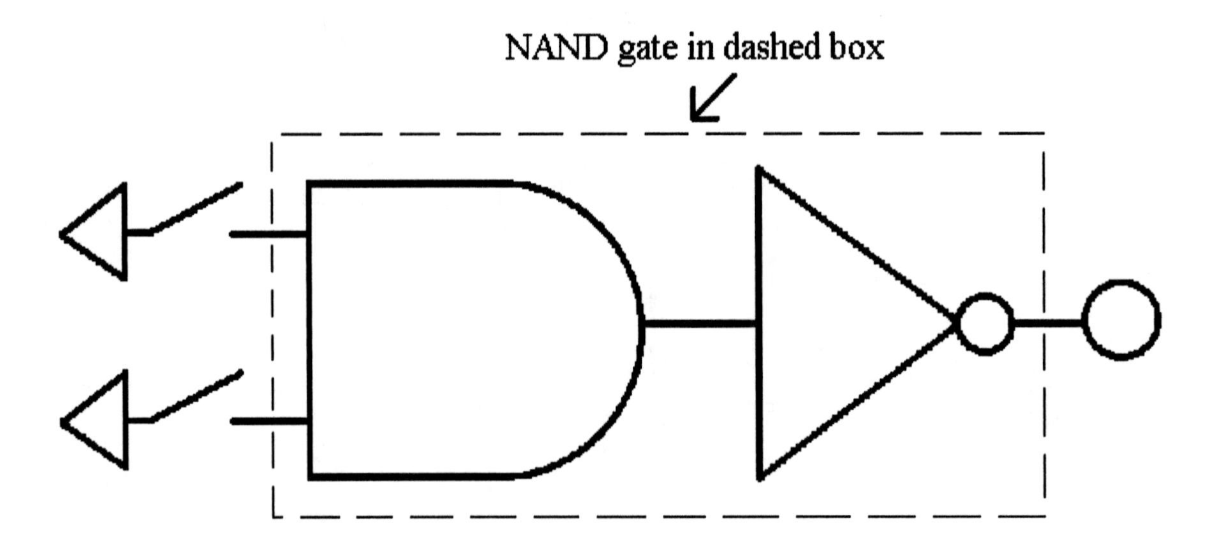

A NAND gate can be represented by the single symbol in the circuit below.

NAND Gate Circuit

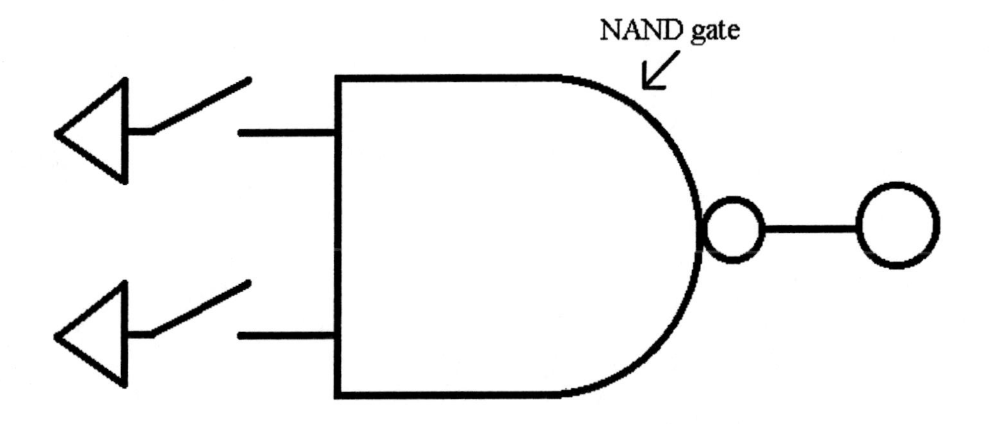

A lone NAND gate is pictured below.

NAND Gate

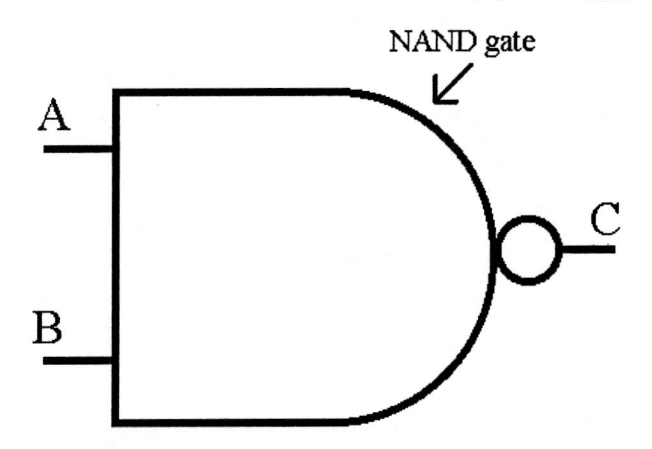

The truth table for the NAND gate is shown below.

NAND gate truth table		
A	B	C
0	0	1
0	1	1
1	0	1
1	1	0

MEMORY

(Address) Decoder

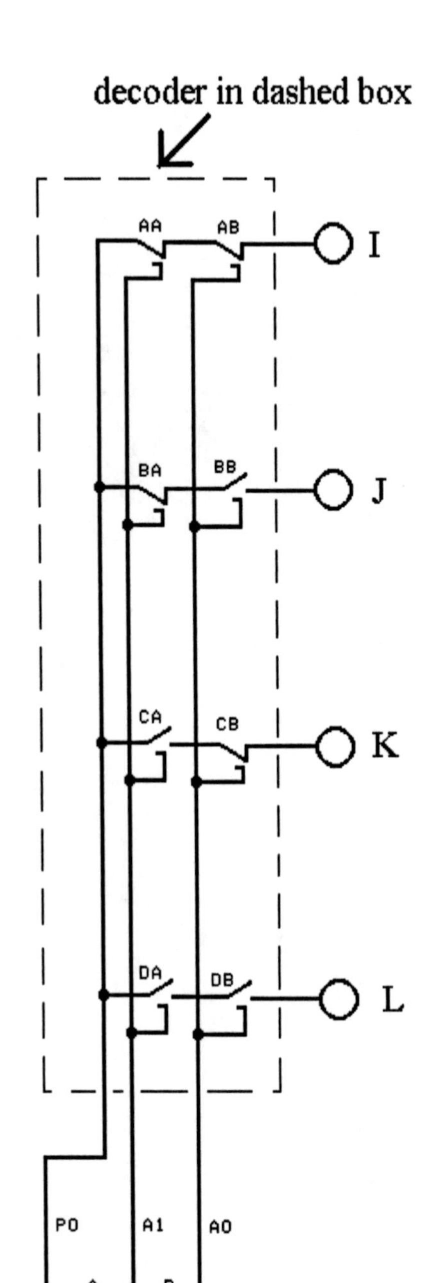

The diagram at left shows a 'decoder.' A and B are the inputs to the decoder and I, J, K, and L are the outputs. The truth table for this circuit is shown below.

A	B	I	J	K	L
0	0	1	0	0	0
0	1	0	1	0	0
1	0	0	0	1	0
1	1	0	0	0	1

Normally closed relay AA is closed. Normally closed relay AB is also closed. Therefore, electricity can travel from the top of the battery, through AA and AB, to light I.

If keys A and B are both pressed, then normally open relays DA and DB are closed (because their electromagnets are powered) and electricity can reach light L.

Similarly, if key A is pressed and key B is *not* pressed, then normally open relay CA is closed and *normally closed relay* CB is closed and light K is on.

Finally, if key A is *not* pressed and key B is pressed, then light J is on.

Wire PO is power. A1 and A0 are address wire 1 and address wire 0. PO has value 1. A1 can have value 1 or 0, and A0 can have value 1 or 0.

Truth Table Generator

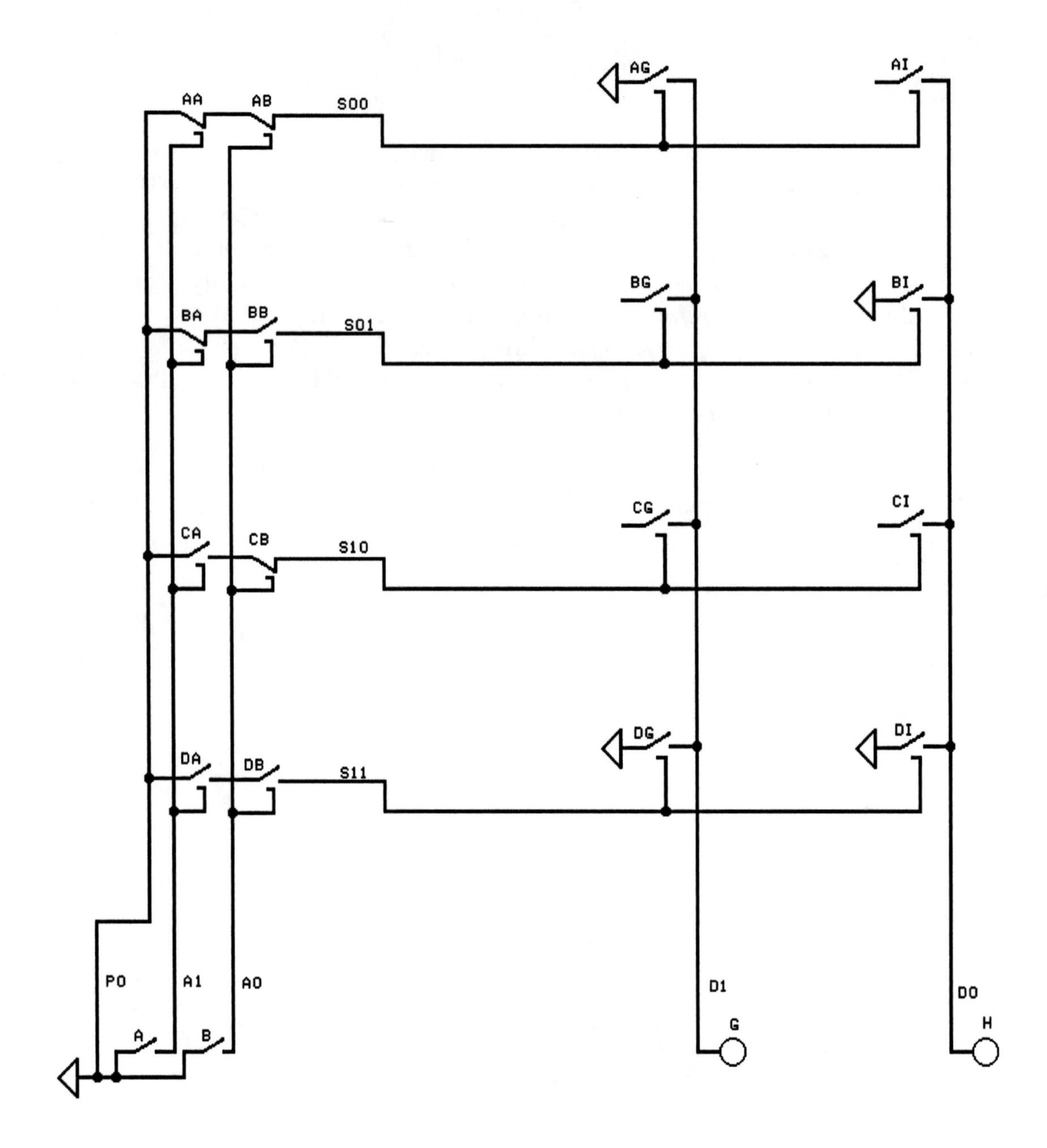

In the preceding circuit, keys A and B are the inputs and lights G and H are the outputs. The truth table for the circuit above is shown below.

A	B	G	H
0	0	1	0
0	1	0	1
1	0	0	0
1	1	1	1

For example, if neither A nor B is pressed, then S00 is powered (has value 1) because the normally closed relays AA and AB are then closed. BB is open so S01 is 0, CA is open so S10 is 0, and both DA and DB are open so S11 is 0. Because S00 is powered, AG is closed and electricity can go from the top of the battery (indicated by a triangle), through relay AG, to wire D1 to light G, so G is on. Relay AI is also closed but relay AI's key is *not* connected to the top of the battery so *no* electricity gets to light H.

For another example, if *both* keys A and B are pressed, then A=1 and B=1 and relays DA and DB are closed. That makes S11=1 and closes relays DG and DI. Electricity can go from the top of the battery through DG and D1 to light G and through DI and D0 to light H. Therefore, A=1 and B=1 results in G=1 and H=1 as in the truth table.

D1 and D0 are data wire 1 and data wire 0. D1 can have value 1 or 0 and D0 can be 1 or 0.

ROM (Read-Only Memory) With Enable (EN) Key (D)

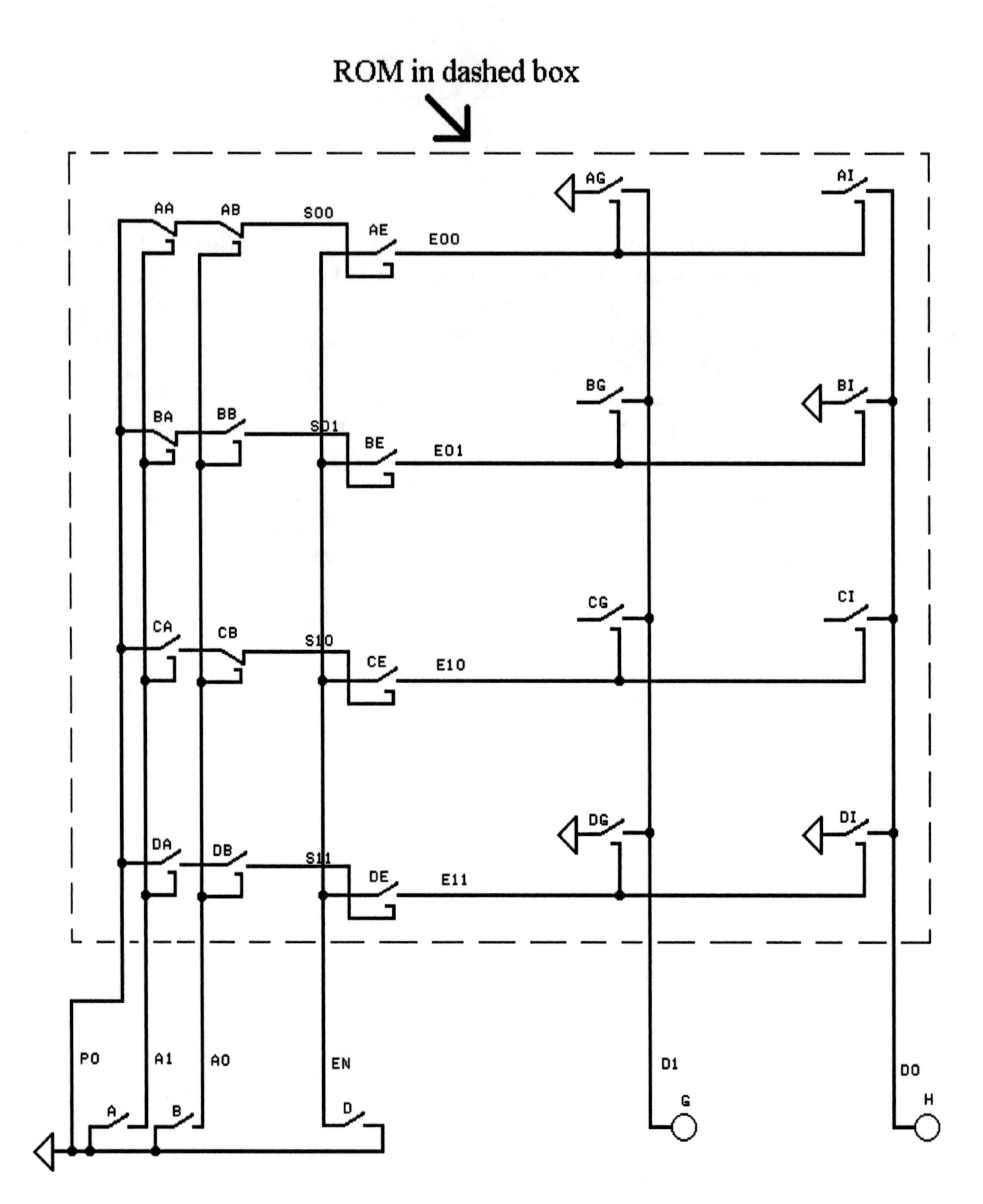

The circuit above has the following truth table:

EN	A1	A0	D1	D0
0	0	0	0	0
0	0	1	0	0
0	1	0	0	0
0	1	1	0	0
1	0	0	1	0
1	0	1	0	1
1	1	0	0	0
1	1	1	1	1

If key D (EN) is *not* pressed ('EN' stands for 'enable.'), then EN is 0, so *no* electricity gets to the electromagnets of AG and AI. Similarly, BG, BI, CG, CI, DG, and DI are open if D (EN) is *not* pressed. Therefore, if D (EN) is *not* pressed, then *no* electricity can get to lights G and H as indicated in the truth table.

If A and B are *not* pressed (A1=0 and A0=0), then electricity gets to the electromagnet of AE and closes relay AE. If D is then pressed (EN=1), then electricity can go from the top of the battery, through D and through AE to the electromagnets of AG and AI. AG and AI then close and electricity can go from the top of the battery, through AG, to wire D1 and light G.

The truth table above can also be represented as below.

EN	A1	A0	D1	D0
0	X	X	0	0
1	0	0	1	0
1	0	1	0	1
1	1	0	0	0
1	1	1	1	1

The X's mean 0 *or* 1. That is, the row with X's means that if EN is 0, then D1=0 and D0=0 no matter what values A1 and A0 have.

Loops Added

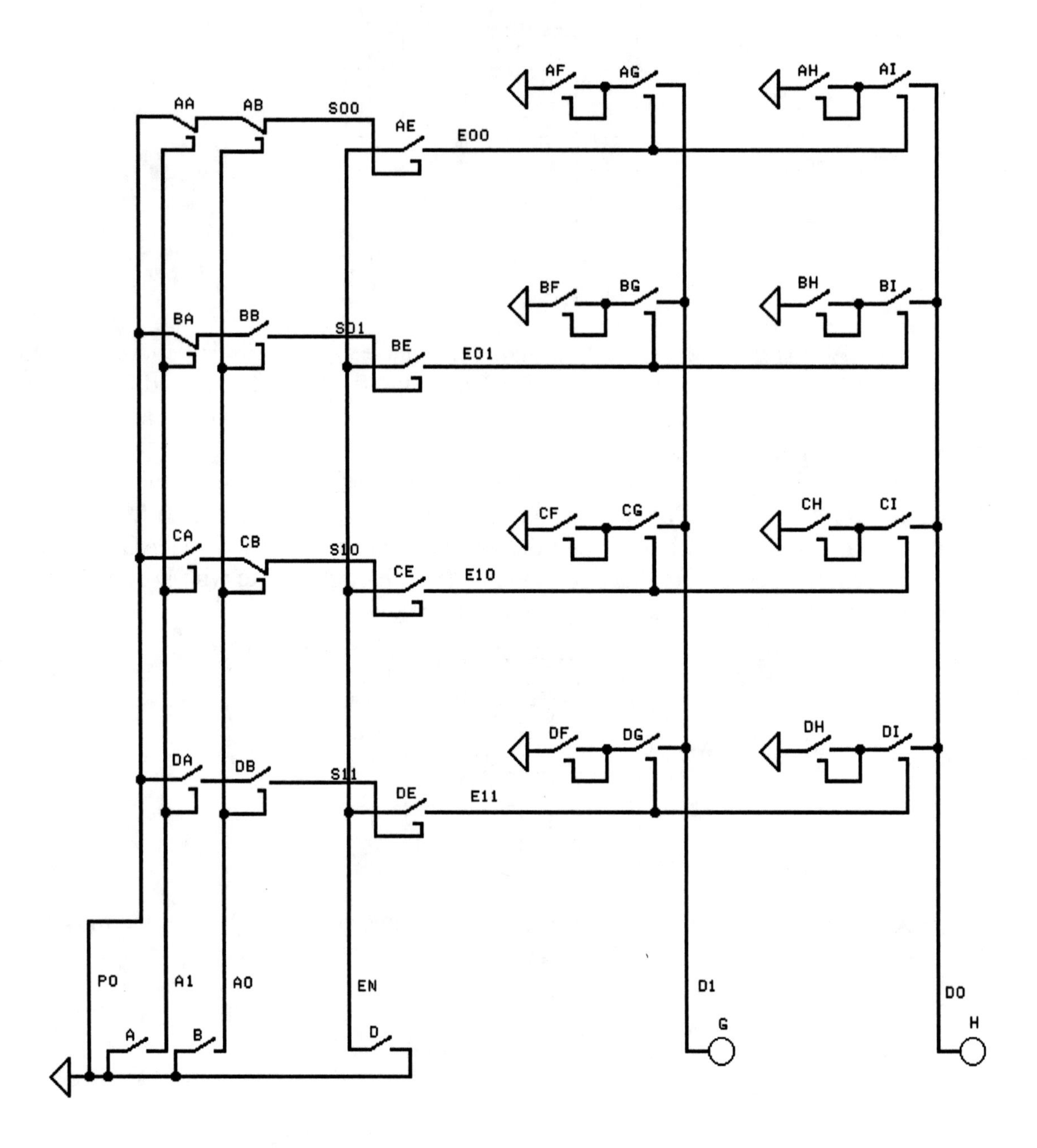

AF
AG
AH
AI
AA
AB
S00
AE
E00
BF
BG
BH
BI
BA
BB
S01
BE
E01
CF
CG
CH
CI
CA
CB
S10
CE
E10
DF
DG
DH
DI
DA
DB
S11
DE
E11
P0
A1
A0
EN
D1
D0
A
B
D
G
H

In the circuit above, eight loops have been added to the previous circuit. The loops are labeled AF, AH, BF, BH, CF, CH, DF, and DH. Each loop can have value 0 or 1. The truth table for this circuit is shown below.

EN	A1	A0	D1	D0
0	X	X	0	0
1	0	0	AF	AH
1	0	1	BF	BH
1	1	0	CF	CH
1	1	1	DF	DH

To make loop AF have value 1, just press key AF down. Key AF will stay down because it is part of a loop. To make AF have value 0 again, just lift key AF up. It will stay up on its own. In the truth table, 'AF' means the value of loop AF. The other loops, AH, BF, BH, CF, CH, DF, and DH, operate similarly.

Input Keys Added

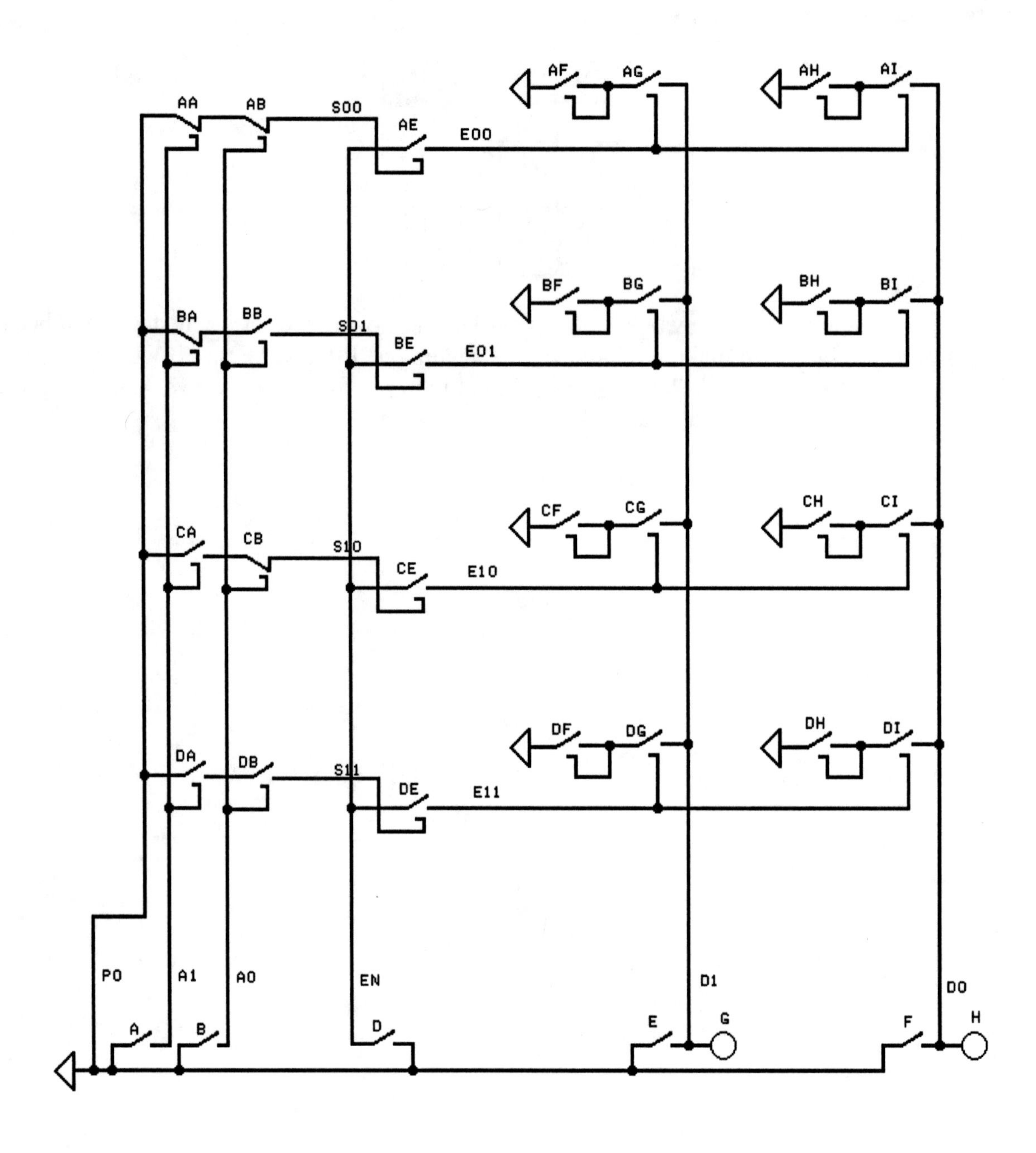

In the diagram at left, key E and key F (bottom right) have been added to the circuit. Keys E and F allow one to set a loop to value 1 without touching the loop's key.

For example, to set loop AF to 1 without touching key AF, one must *not* push key A or key B, which closes relay AE. Then you hold down key E to put value 1 on wire D1. Finally, temporarily pushing key D makes EN temporarily 1. Because AE is closed, EN=1 powers relay AG's electromagnet and closes AG. D1's value of 1 can now go through key AG to loop AF, thereby making loop AF have value 1.

Memory (Clear Key Added)

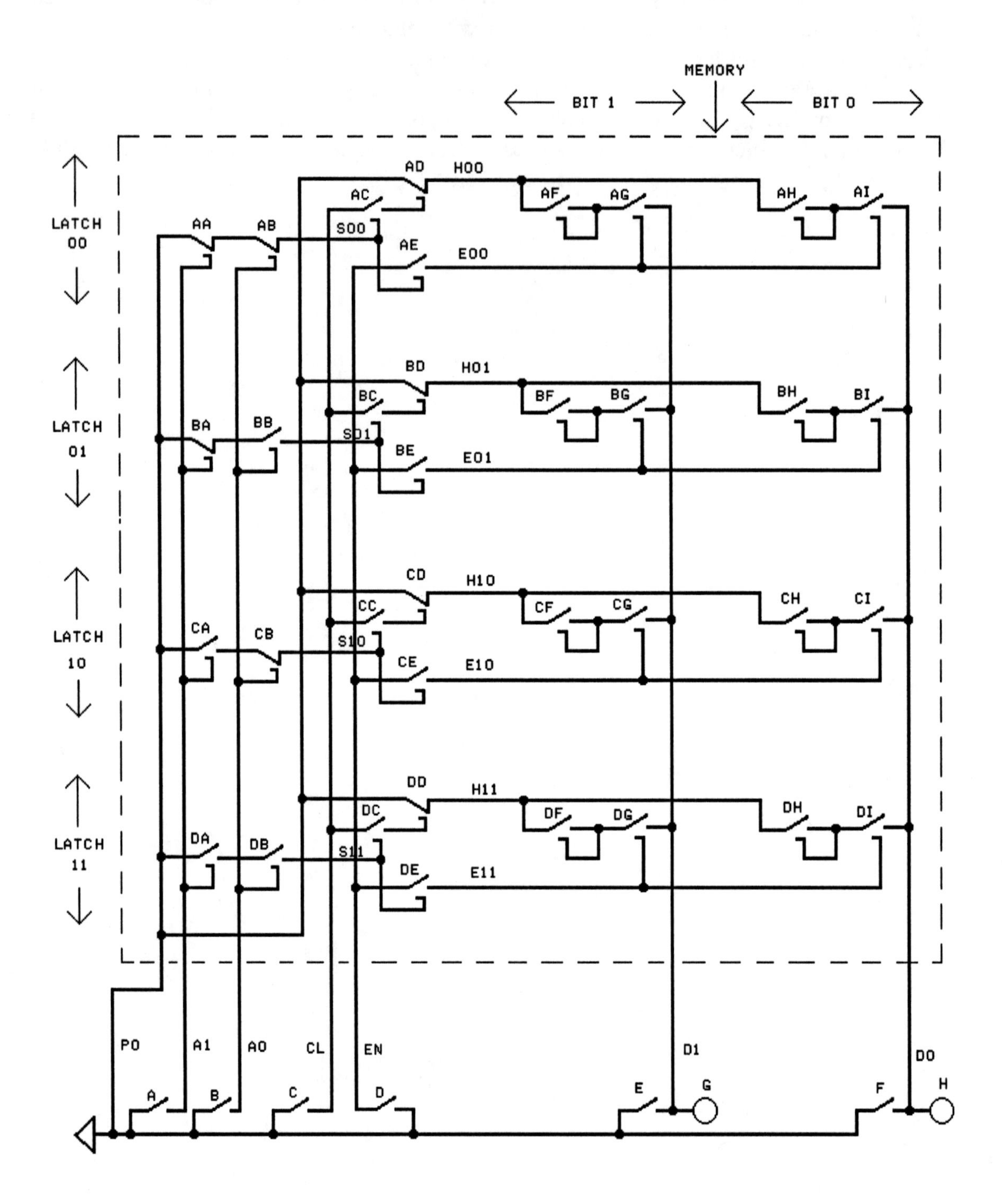

In this circuit, key C, wire CL (for CLear), relays AC, BC, CC, and DC, normally closed relays AD, BD, CD, and DD, and wires H00, H01, H10 and H11 (H for Hold, or remember) have been added to the circuit. These additions allow loops to be 'cleared' to value 0 by manipulating keys outside the dashed box (memory) without touching the loop keys.

The diagram above shows a *memory* within the dashed box. The memory can be controlled by the keys outside the dashed box at the bottom of the diagram. What a memory does will be explained first. Then, *how* the memory works will be described.

AF, AH, BF, BH, CF, CH, DF, and DH are each relay keys of loops. You can change the value of loop AF from 0 to 1 by simply pressing key AF down. Similarly, you can change the value of loop AF from 1 to 0 by lifting key AF. To determine whether a loop has value 0 or value 1, just look at the loop's key. If the key is down, then the loop has value 1. If the key is up, then the loop has value 0. The value of a loop stays the same until you change it.

However, suppose that the dashed box was a physical box and you could not reach inside the box. If you buy a memory chip at a store, the circuitry is enclosed in a plastic box with wires PO, A1, A0, CL (This may be called WR for 'WRite.'), EN, D1, D0 and *GRound wire*, GR, sticking out. The circuitry uses transistors instead of relays for switches, so even if you broke the box open, you couldn't change the values by hand. (A memory from a store would probably have more address lines (wires) like A2, A3,...A20 and data lines like D2, D3,...D7.)

The memory is constructed so that the values in the loops can be examined and changed using only keys A, B, C, D, E, and F and light bulbs G and H which are all outside the box and are *not* part of the memory.

Where Power Reaches in a Memory

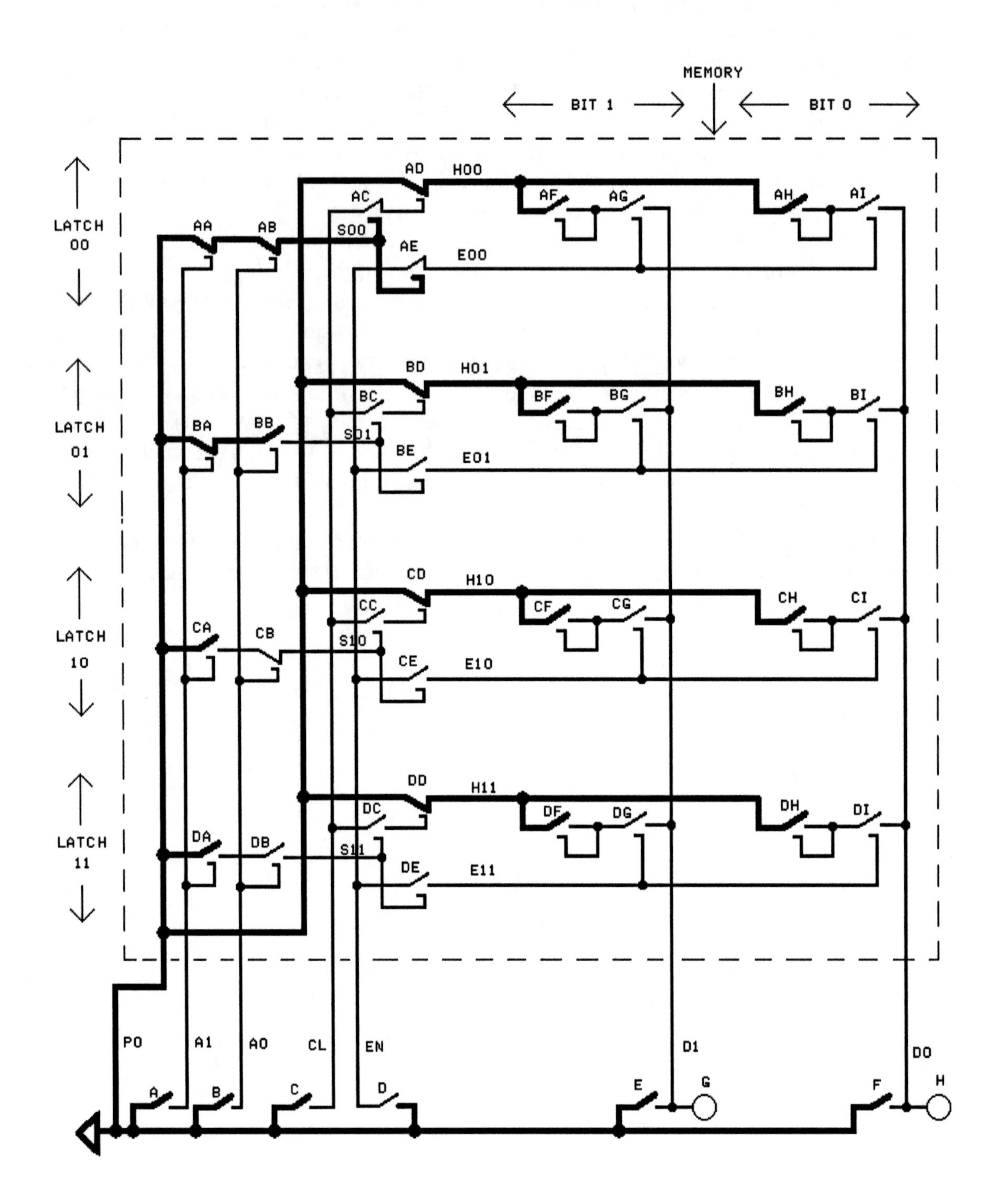

*The **bold** wires in the diagram at left, show which wires are powered.*

A wire is powered only if it is connected to the top of the battery (represented by a triangle in the lower left corner of the diagram, as shown below).

power

Notice the new symbol used for keys AC and AE. Keys AC and AE are *normally open keys*. However, they are *closed now* because their electromagnets are powered. Therefore, they are represented as:

***closed*, normally open key**

instead of as:

open, normally open key

Electricity can flow from left to right (or right to left) through a closed key even if it's a closed but *normally open key*.

Similarly, an *open*, normally closed key is represented as:

***open*, normally closed key**

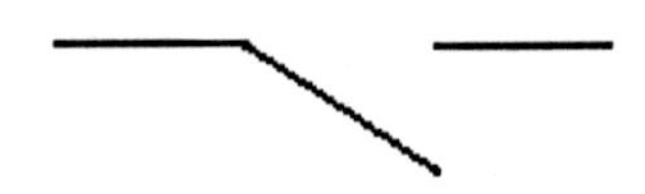

instead of as:

closed, normally closed key

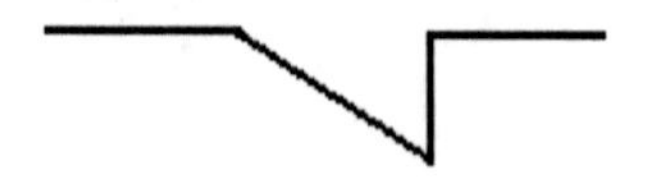

Notice that, in the diagram of memory, all of the loops (AF, AH, BF, BH, CF, CH, DF, and DH) have value *0* because all of those normally open keys are *open.*

AF is 'bit 1' of 'latch 00' and has *value* 0. AH is bit 0 of latch 00 and also has *value* 0.

You should follow the power from the top of the battery (the triangle in the lower left of the diagram of memory *above*) and see why certain wires are *bold* and the rest are normal. Remember, electricity *can't* go through *open* keys. Electricity also does *not* travel between *crossing wires.* Crossing wires are *not* touching (not connected). You should also understand why some electromagnets are powered and others aren't, and how powering the electromagnet of a key *closes* a *normally open key* and, in later diagrams, *opens* a *normally closed key.*

Now, suppose we want to store value 01 in latch 10. This means we want to keep key CF *open* for value 0 and *close* key CH for value 1. This is called '*writing*' value 01 to *address* 10.

To do this, you *first* select latch 10 by pressing key A and *not* pressing key B. This selects latch 10 as indicated by the *bold* select 10 wire, 'S10,' in the diagram *below.* Key A controls address wire 1, labeled A1 in the diagram, and key B controls address wire 0, labeled A0 in the diagram. Both address wires, A1 and A0, together, are called the address bus. A group of similar wires are, together, called a 'bus.' Pressing key A and *not* pressing key B results in power going through the circuit as indicated by *bold lines* in the diagram below. Notice that horizontal wire S10 has power (is bold) while S00, S01, and S11 do *not* have power. This selects latch 10.

Selecting the Address

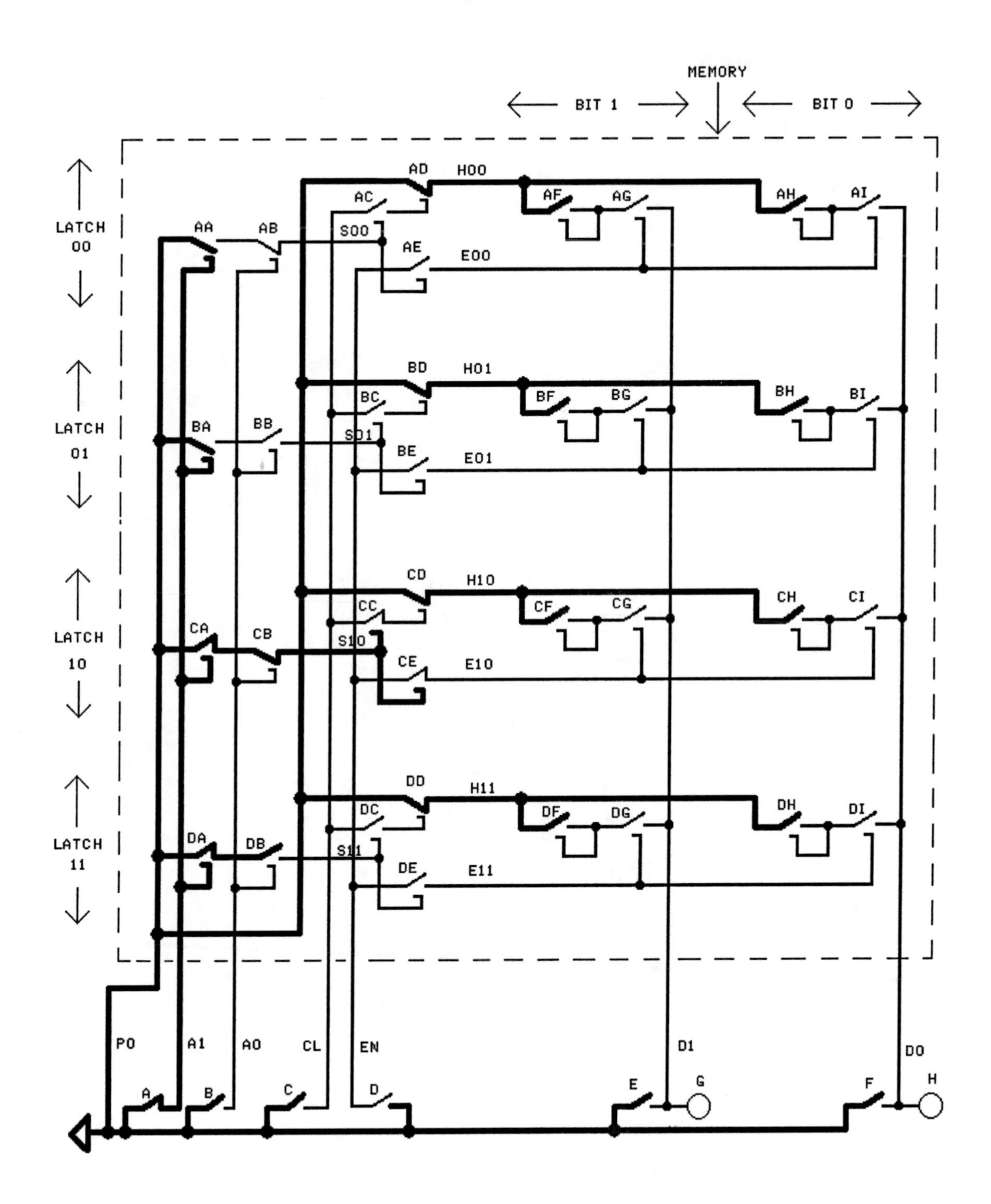

The *second* step in writing value 01 to address (latch) 10 is to press key F and *not* press key E as in the following diagram. *Not* pressing key E gives value 0 to data wire D1 and *pressing* key F gives value 1 to data wire D0. Both data wires, D1 and D0, are, together, called the 'data bus' just as both address wires, A1 and A0, are, together, called the 'address bus.' The first and second steps can be done simultaneously. This results in power going through the circuit as indicated by the *bold* wires in the diagram below. Wire D0 is bold and, so, has value 1.

Selecting the Data to be Written

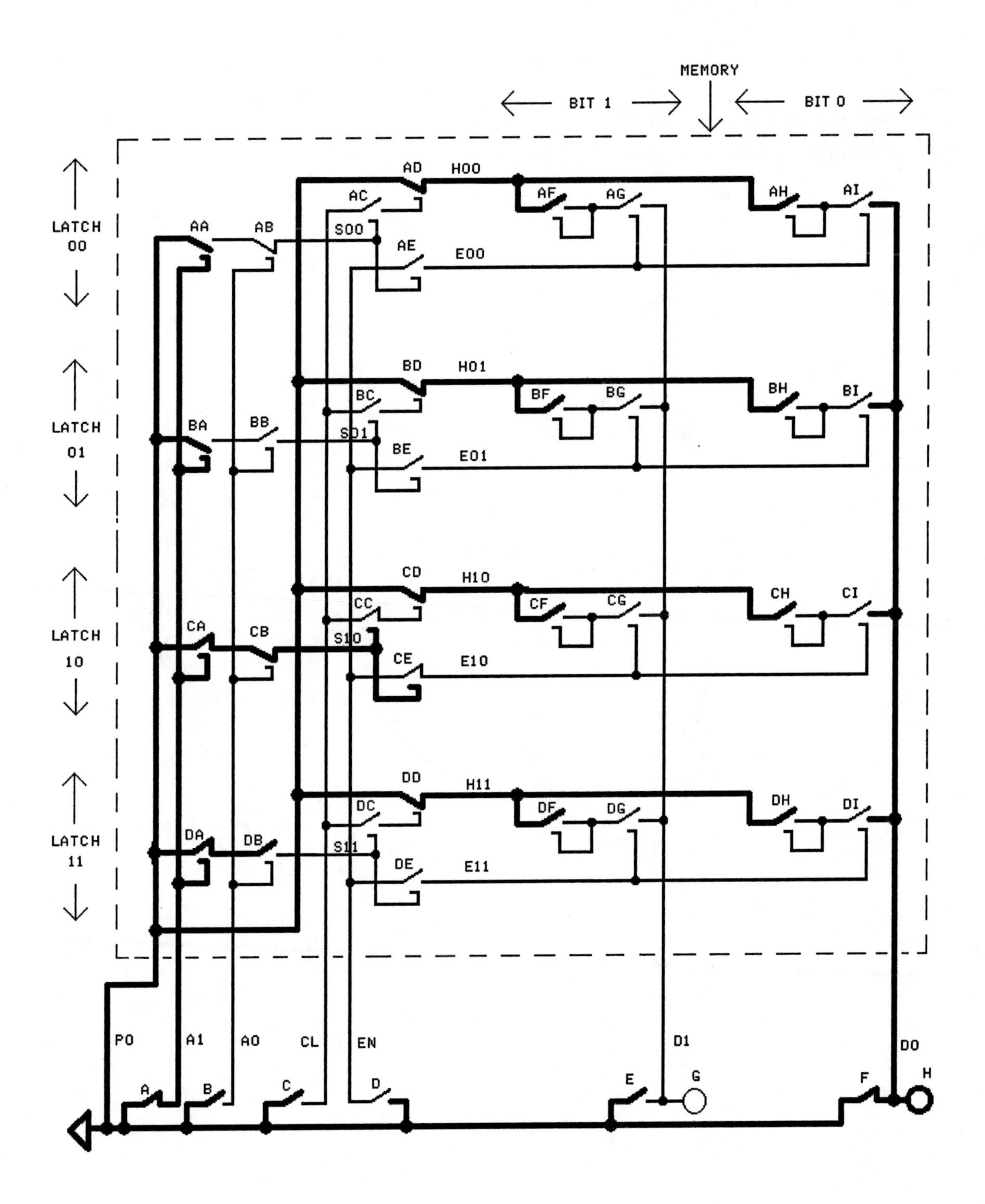

The *third* step in writing 01 to address 10 is pressing the enable key, 'D,' which controls the enable ('EN') wire. This results in power going through the circuit as indicated in bold in the following diagram. Notice that *loop* CH now has value 1. Loop CH got power from wire D0 through CI. *No* power went from wire D1 through CG to loop CF because wire D1 is *not* powered.

It's important to remember that pressing the enable key, 'D,' makes EN=1 and connects the loops of the selected (by the address wires A1 and A0) latch to the data bus wires, D1 and D0.

Pressing Enable (EN)

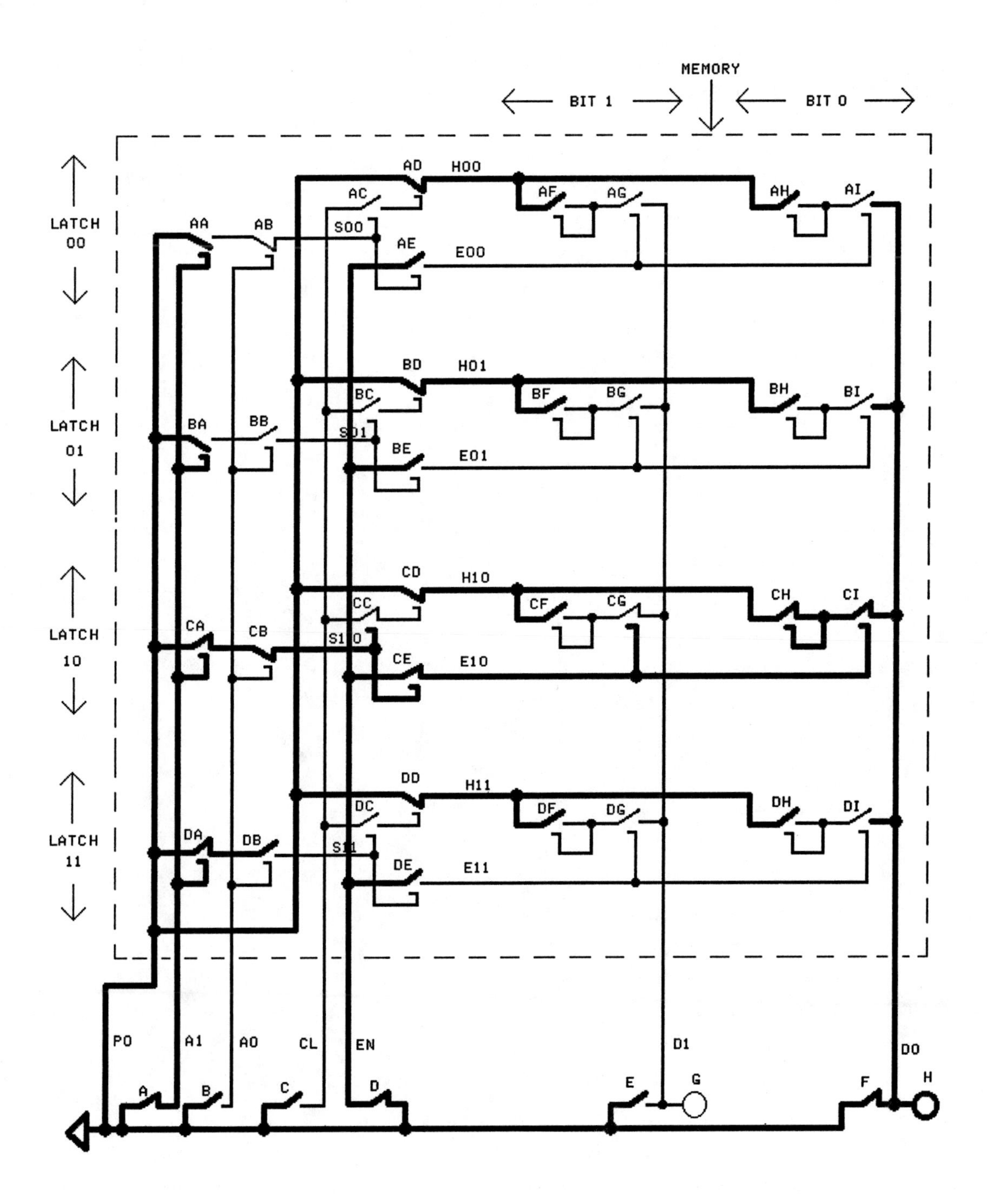

In the fourth step, key D is released and the enable (EN) wire returns to value 0 (unpowered). This results in power flowing through the memory as indicated by bold wires in the following diagram. Notice that loop CH still has value 1 even though loop CH is no longer connected to data wire D0 through relay CI (because relay CI is open).

Releasing Enable (EN)

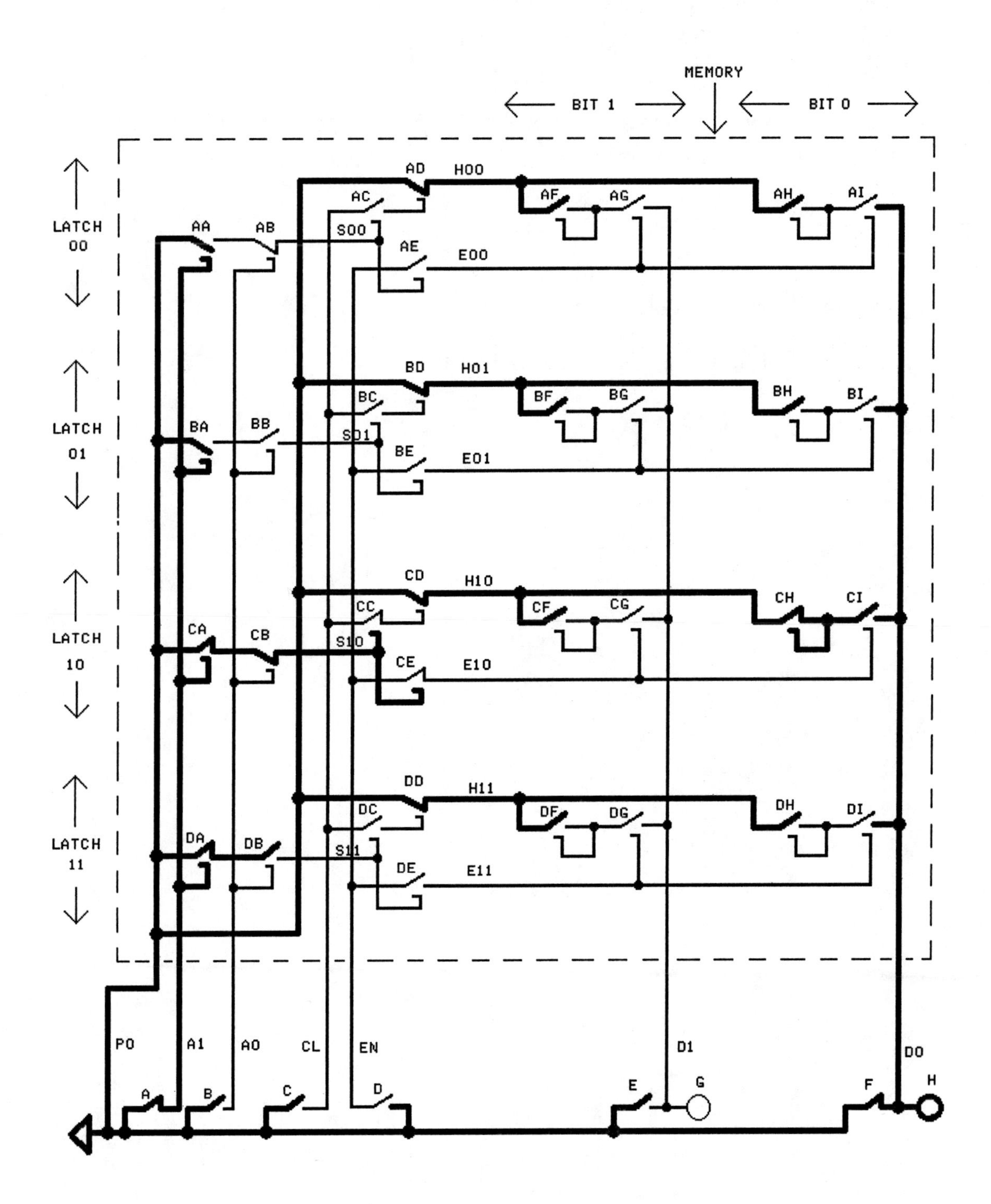

Step five: Keys A and F are released and address *wire* A0 and data *wire* D0 get value 0 (as indicated in the following diagram). Notice that *loop CH still has value 1.*

Therefore, to write value 01 to latch 10, you press A and *not* B to select latch 10; and, to choose data 01, do *not* push E and push F. This makes wire A1 have value 1, wire A0 have value 0, wire D1 have value 0, and wire D0 have value 1. Then, while holding A and F down, temporarily press D to make the enable wire, EN, temporarily 1. Then, release A and F. This can be described as follows.

1. Select the address and data with the address and data keys A, B, E, and F (A1, A0, D1 and D0).
2. Temporarily press D (EN).
3. Release the address and data keys.

That's all there is to storing data in an *empty* latch.

Releasing the Address and Data Keys

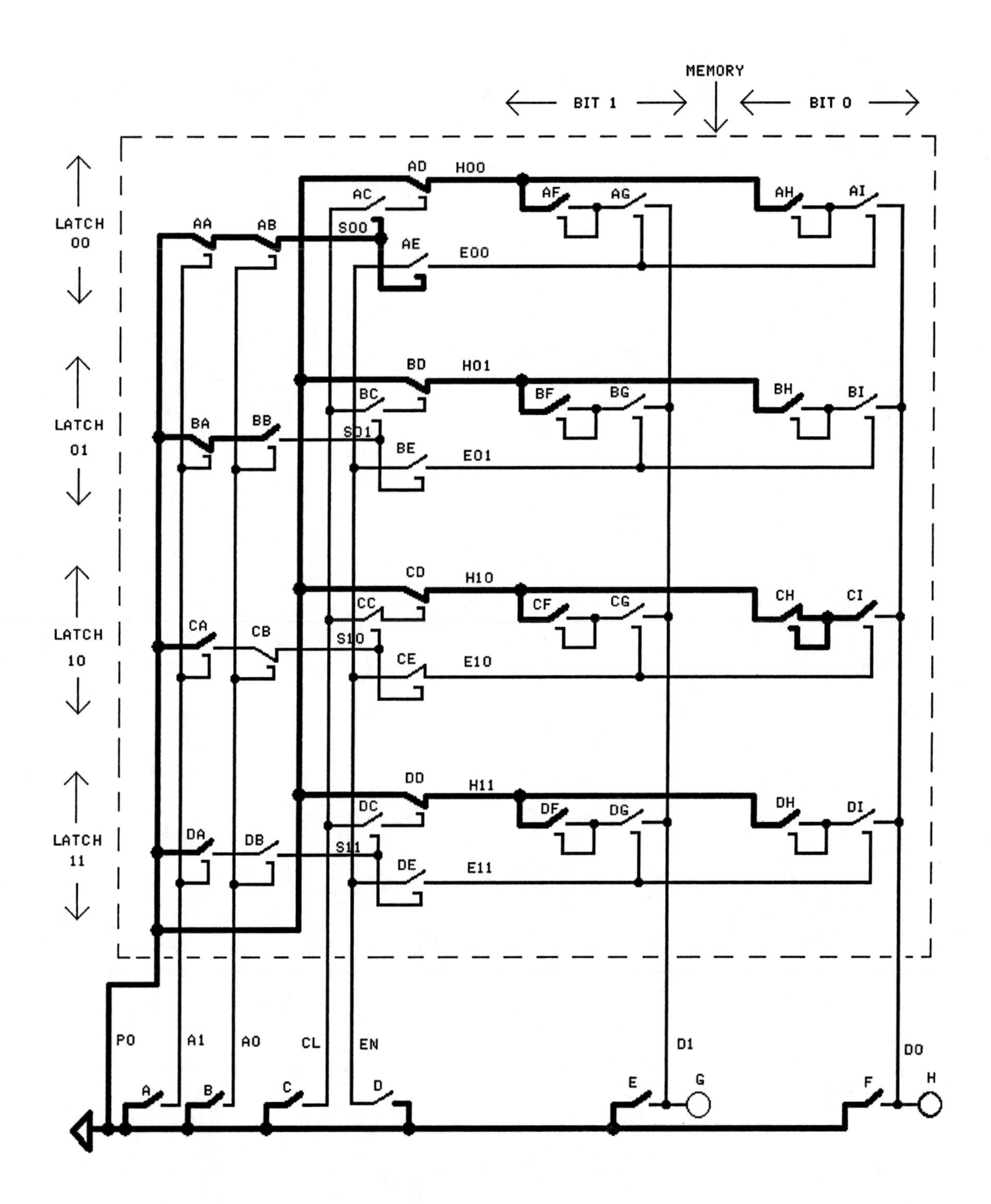

To find out what is in a latch, do the following.

1. Select the address of the latch you want to read with keys A (wire A1) and B (wire A0).
2. Press key 'D' to make the 'EN' wire have value 1. The lights G and H will indicate the values of the data bits stored in that latch.
3. Release the enable key D.
4. Release the address keys, A and B.

For example, to read latch 10, *first* press key A (and *not* key B) to select latch 10 (as indicated in the following diagram).

Selecting the Address to Read

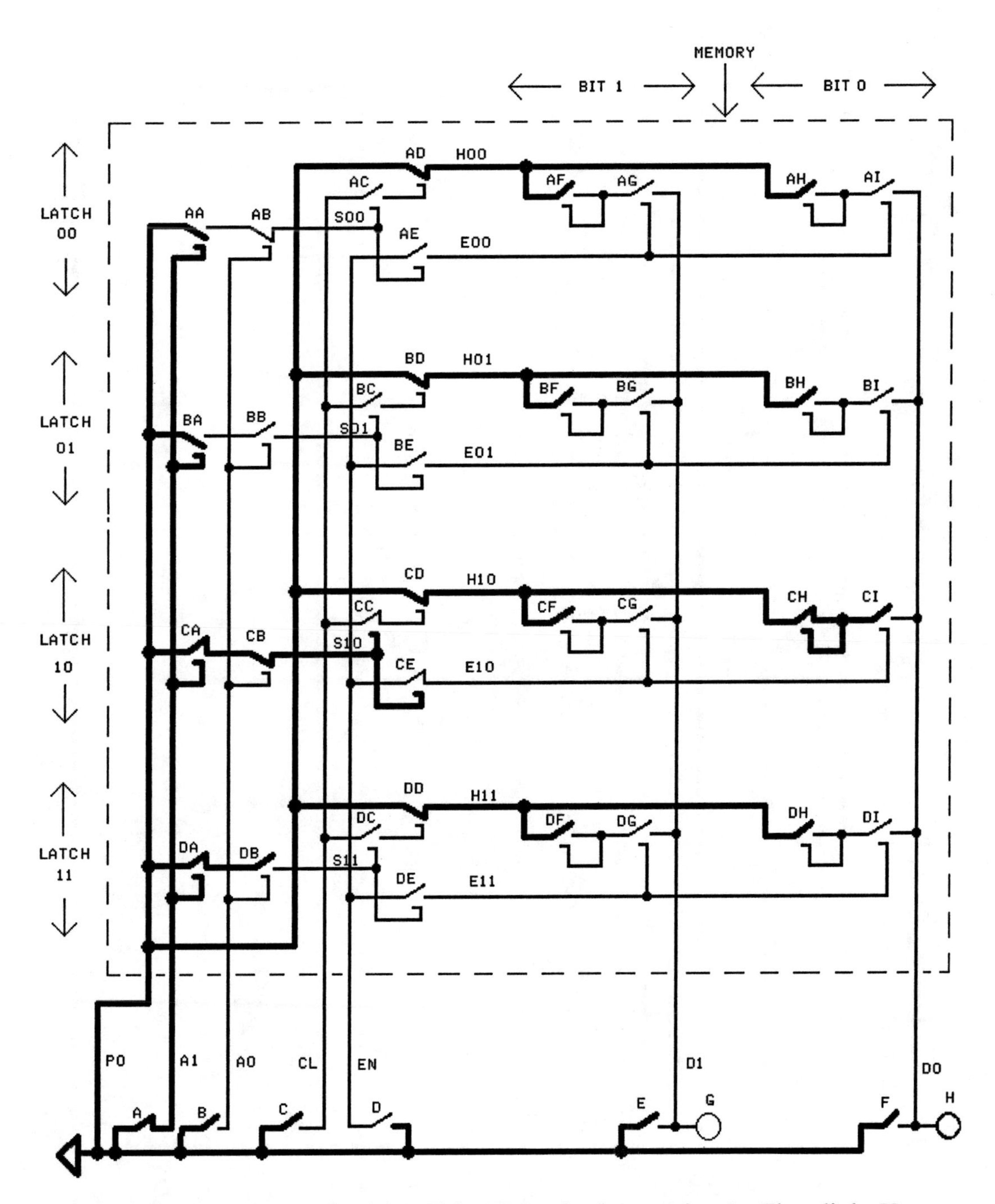

Second, press key D to make the enable (EN) wire have value 1. Then light H comes on indicating that bit 0 (loop CH) of latch 10 has value 1 and light G stays off indicating that latch 10 has value 0 in bit 1 (loop CF). This is shown in the following diagram.

Notice that making wire EN have value 1 connects the loops of the selected latch to the data wires D1 and D0.

Enabling (EN) the Output

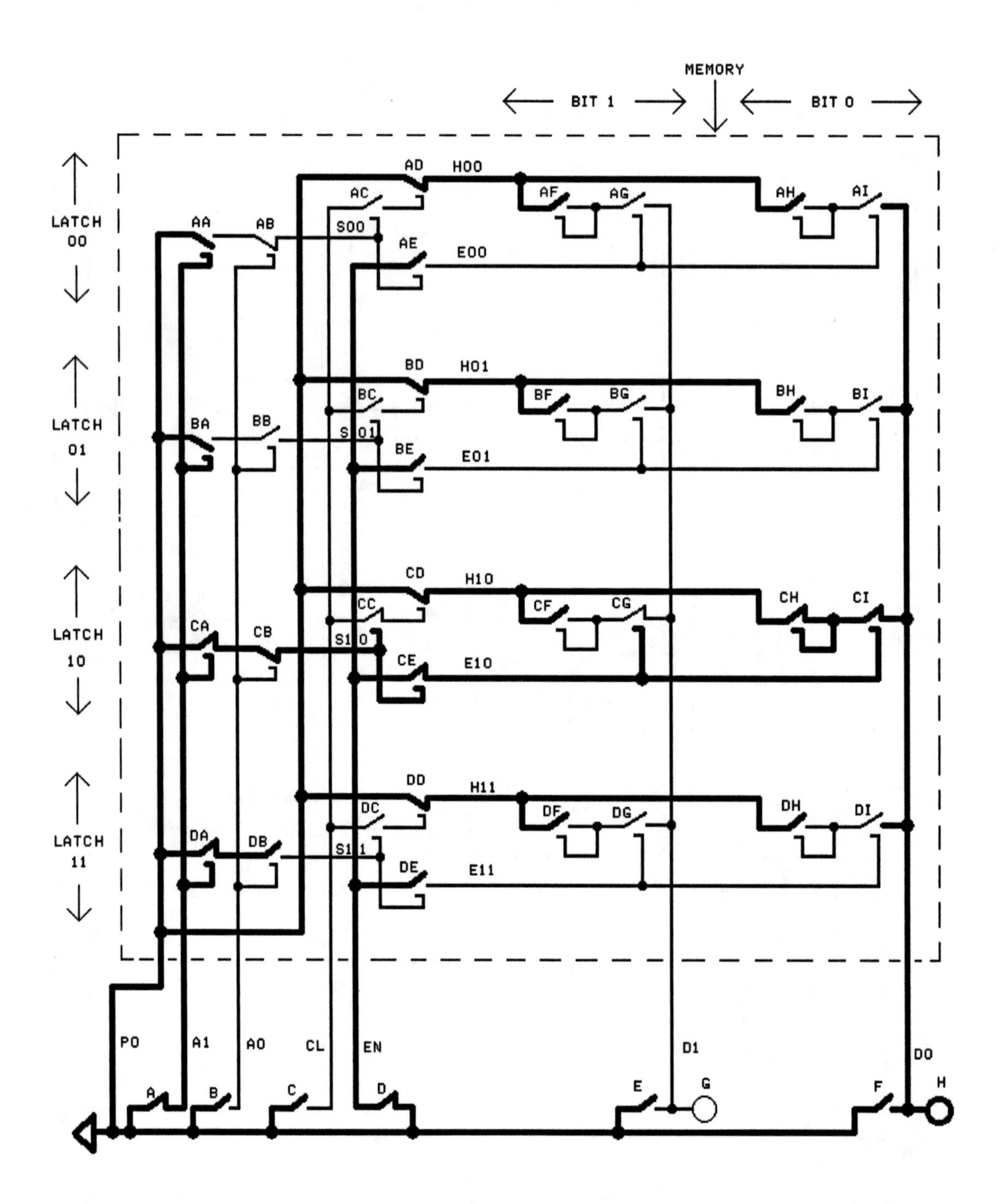

Third, release key D when done reading latch 10 (as indicated in the following diagram). That's all there is to reading a latch in memory.

Releasing Enable (EN)

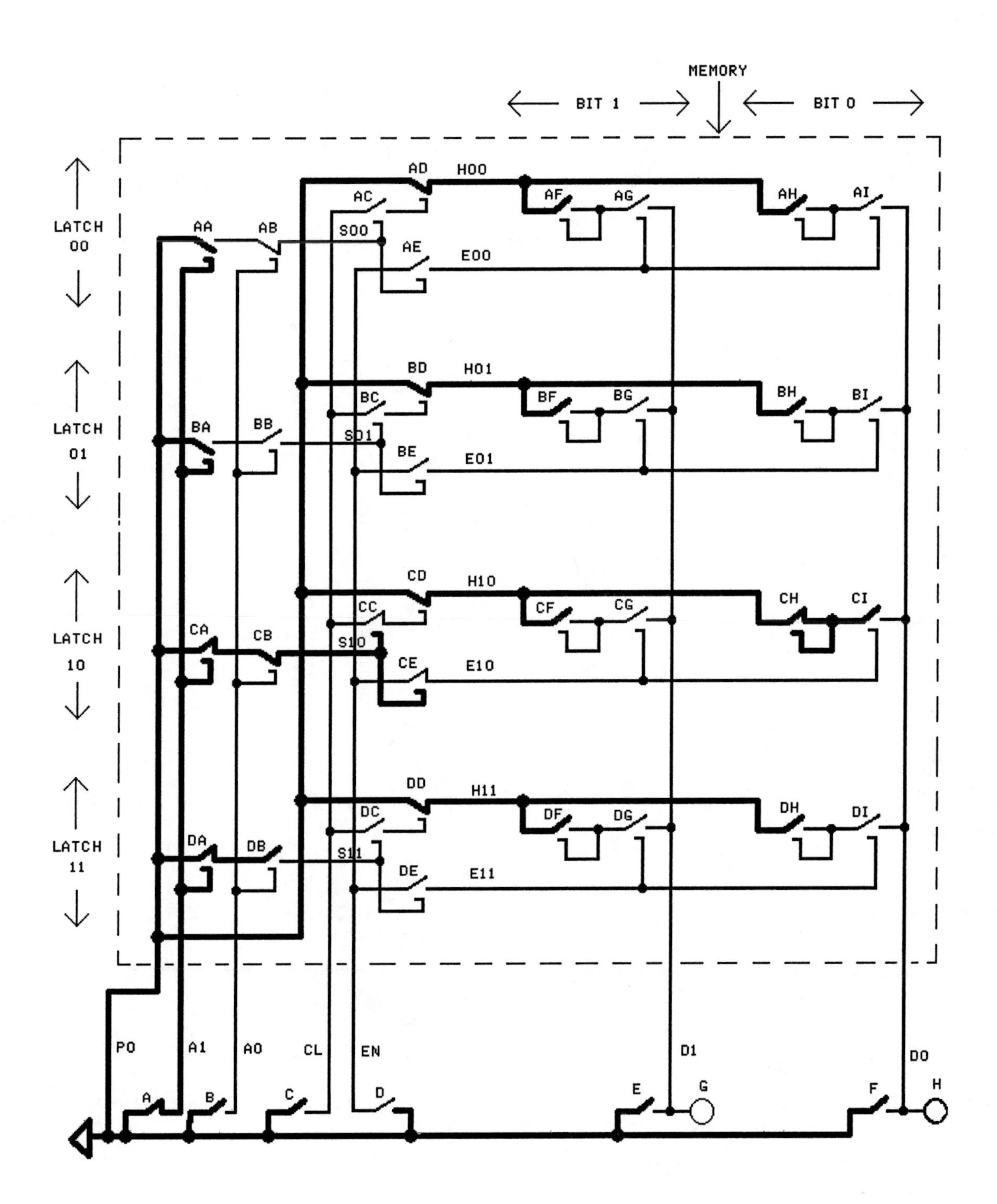

To *erase* a value from a latch and make all of the latch's loops have value 0, do the following.

1. Select the latch with keys A (A1) and B (A0).
2. Temporarily press key C to make the 'clear' (CL) wire temporarily have value 1.

The following diagram shows latch 10 selected by pressing A and *not* pressing key B. It also shows key C being pressed to clear both of latch 10's data bits to 0. Don't press C until after A is pressed (so that no other latch is accidentally erased).

Notice that pressing C makes the selected 'H' (for 'Hold') wire, 'H10,' have value 0.

Selecting the Address and Clearing (with CL)

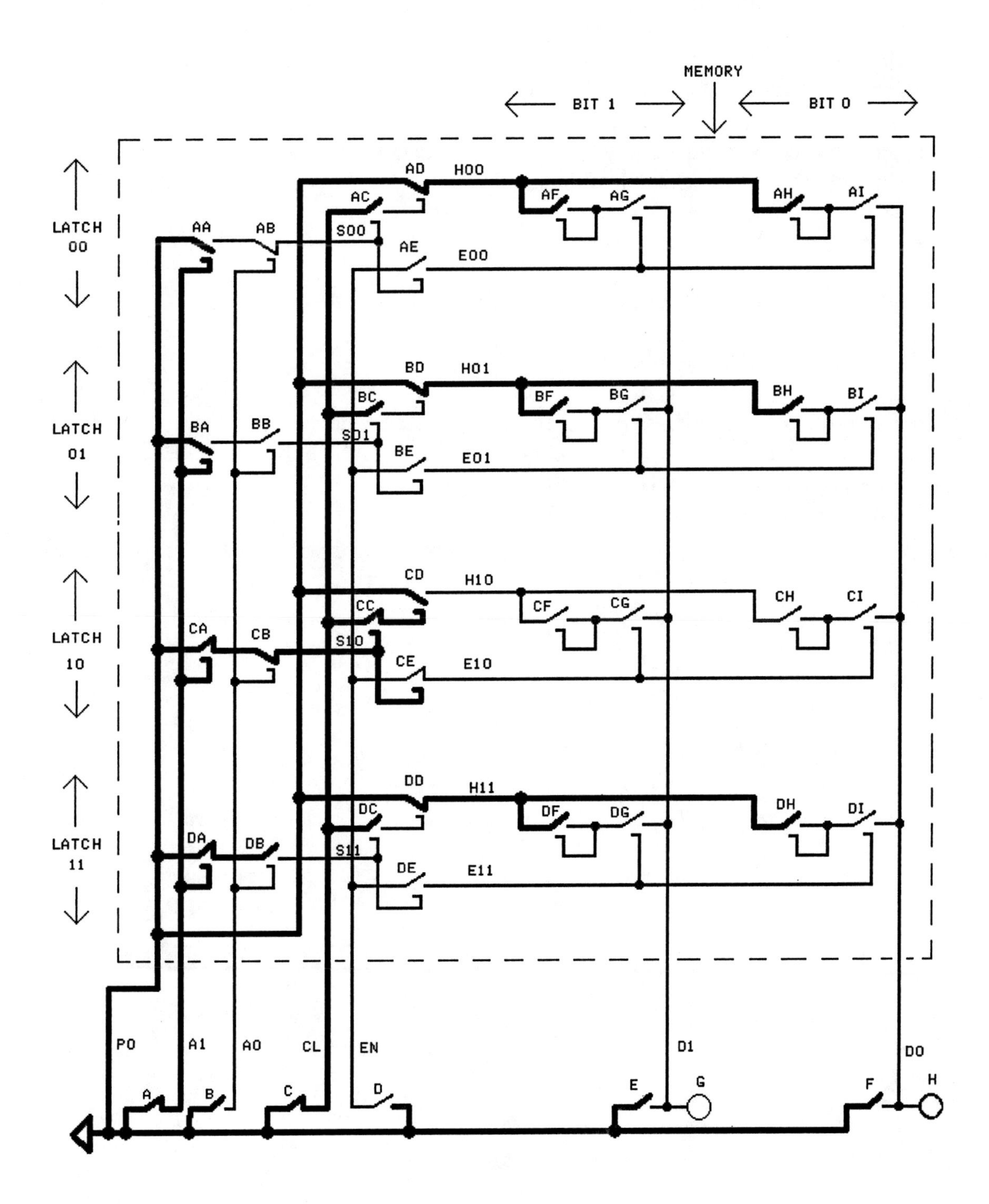

Release key CL after clearing latch 10 as indicated in the following diagram. Don't release key A until *after* CL is released so that you don't accidentally erase another latch.

Writing to a latch will *not* clear *any* bits that were previously 1, so *always* clear a latch before copying (writing) data to it. Therefore, to write to a latch, do the following.

1. Press the correct address keys (A and B) *and* data keys (E and F).
2. Press the clear key, C, to clear the latch.
3. Release the clear key, C.
4. Press the enable key, D, to send data from the data wires (D1 and D0) to the latch.
5. Release the enable key, D.
6. Release the address keys (A and B) and the data keys (E and F).

To *read* data, just do the following.

1. Press the correct address keys, A and B, to select the latch to read.
2. Press the enable key, D, to send the latch's values to the lights, G and H.
3. Release the enable key D.
4. Release the address keys, A and B.

Releasing CL

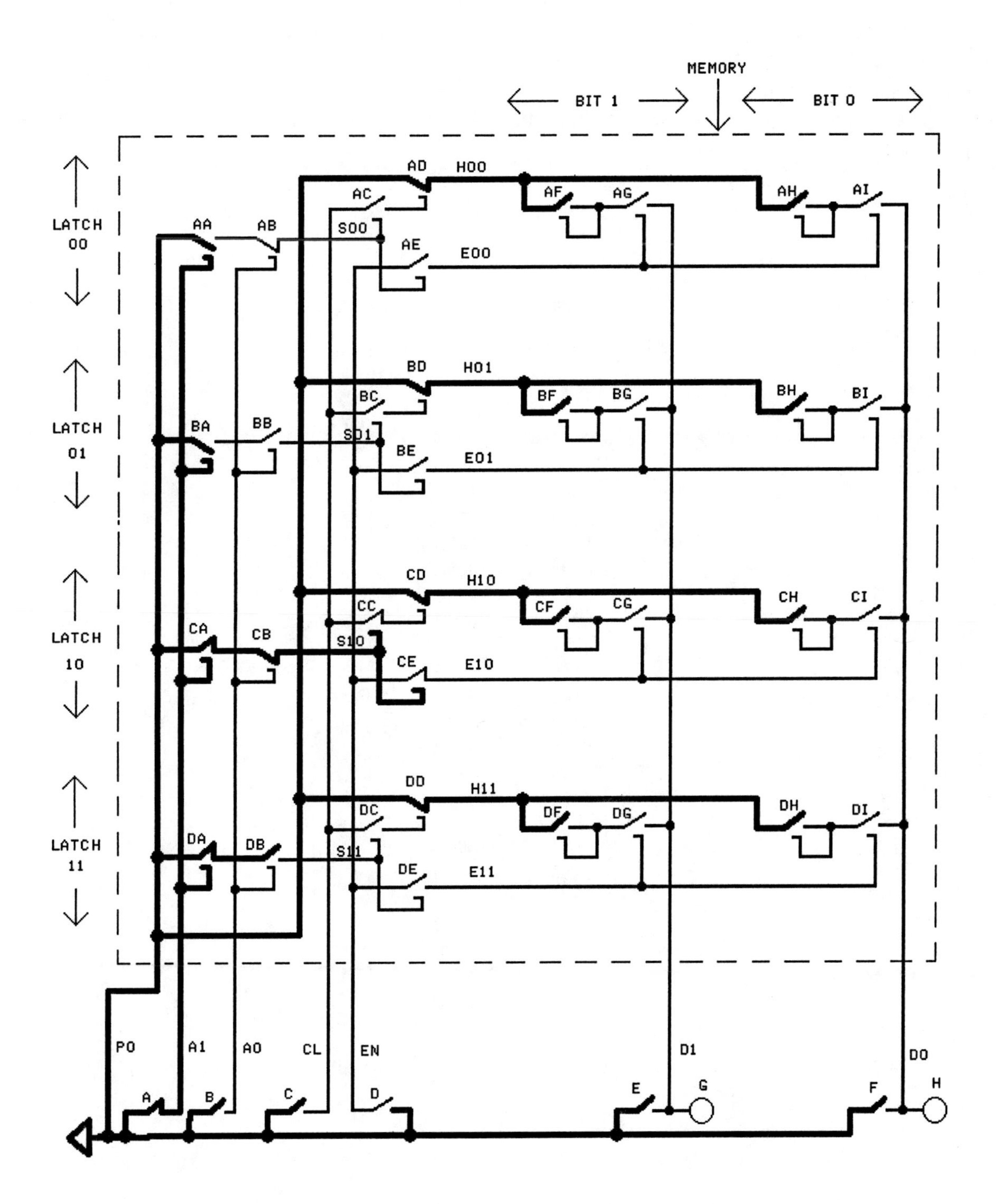

The memory in the drawings is very small. There are only two address wires and two data wires. Because there are two address wires, there are four possible addresses: 00, 01, 10, and 11, and, so, four latches. Because there are two data wires, each latch has two loops. Each loop holds one 'bit' of information, a 0 or a 1. Four latches with two loops each means 8 (= 4 x 2) loops total.

The table below shows all bit values in each latch when data 01 has been written to the loops of latch 10.

latch address	bit values
00	00
01	00
10	01
11	00

A larger memory with *four* address bits and *four* data bits with 0 in all the loops can be represented as below.

latch address	bit values
0000	0000
0001	0000
0010	0000
0011	0000
0100	0000
0101	0000
0110	0000
0111	0000
1000	0000
1001	0000
1010	0000
1011	0000
1100	0000
1101	0000
1110	0000
1111	0000

INSTRUCTIONS

The next most important part of a computer, after memory, is the processor. A processor changes the values in memory as instructed by instructions stored in memory. An instruction is a group of bits (loop values) in memory that tell the processor to do something. A group of instructions that instruct the processor to do some task is called a program.

The simple kind of processor described in this book has only one type of instruction, but that instruction is sufficient to do anything, as will be seen. The instruction is 'copy' (and 'go to'). Each instruction simply copies some bits of data from somewhere in one latch in memory to somewhere else in another latch in memory. The instruction indicates:

1. which latch to copy data (data is bit (loop) values) from
2. which latch to copy data to
3. which bits to change in the copied-to latch
4. how much to rotate the 'from' data before copying some of its bits to the 'to' data latch
5. which latches to get the next instruction from.

In the computer considered here, the number of address bits is the same as the number of data bits.

The explanation of the following example will not be clear at first, but just read through it. Then reread it. It will be clear later.

Consider a four-address-bit, four-data-bit memory with the values (in loops) below:

Example Program

	latch address	bit values	
	0000	0100	address of instruction
	0001	0001	value of a
	0010	0001	not *0* = 1
	0011	0000	not *1* = 0
instruction_1	0100	0001	from address
	0101	1000	to address
	0110	0001	'to' bits to change
	0111	1000	instr.addr.and rot.amount
instruction_2	1000	0010	from address
	1001	0001	to address
	1010	0001	'to' bits to change
	1011	1100	instr.addr.and rot.amount
instruction_3	1100	0000	from address
	1101	0000	to address
	1110	0000	'to' bits to change
	1111	1100	instr.addr.and rot.amount

Only the 1's and 0's are part of the program. The rest is just comments for a person. The latch addresses are just where the program is stored. *The bit values are the program.* '*Instr.addr.and rot.amount' is short for 'next instruction's address and rotate amount.*'

Latch 0000 holds the value 0100 so that the first instruction is in latches 0100, 0101, 0110, and 0111, and is labeled 'instruction_1' in the program. Latch 0000 is special and always holds the address of the next instruction to be executed.

The first word of instruction_1 is in latch 0100 and is 0001. That means that data (bit values) is copied *from* latch 0001 in memory.

The second word of instuction_1 is in latch 0101 and is 1000 and indicates that the data will be copied *to* latch 1000.

The third word of instruction_1 is in latch 0110 and is 0001 and indicates that only the rightmost bit, and *not* the three leftmost bits, of latch 1000 will be changed because only the rightmost bit of 0001 is 1.

The rightmost two bits of latch 0111 are 00 and indicate that the data in latch 0001 will *not* be rotated at all when the data is copied to latch 1000.

The leftmost two bits of latch 0111 are 10 and indicate that the next instruction will be in latches 1000, 1001, 1010, and 1011. That is, instruction_2 will be executed next.

After instruction_1 is executed, the memory has the following bit values.

	latch address	**bit values**	
	0000	**1000**	**address of instruction**
	0001	**000*1***	**value of a**
	0010	**0001**	**not 0 = 1**
	0011	**0000**	**not 1 = 0**
instruction_1	**0100**	**0001**	**from address**
	0101	**1000**	**to address**
	0110	**0001**	**'to' bits to change**
	0111	***1000***	**instr.addr.and rot.amount**
instruction_2	**1000**	**0011**	**from address**
	1001	**0001**	**to address**
	1010	**0001**	**'to' bits to change**
	1011	**1100**	**instr.addr.and rot.amount**
instruction_3	**1100**	**0000**	**from address**
	1101	**0000**	**to address**
	1110	**0000**	**'to' bits to change**
	1111	**1100**	**instr.addr.and rot.amount**

The underlined loop values (bits) were copied to when instruction_1 was executed. The *italics* show from where data was copied. The arrows show how data was copied. The rightmost bit of latch 0001 has been copied to the rightmost bit of latch 1000. Also, all bits of latch 0111 (that is, 1000) have been copied to latch 0000 indicating that the next instruction will be in latches 1000, 1001, 1010, and 1011 (instruction_2). That is, after the first instruction, instruction_1, is executed, latch 0000 has value 1000.

The leftmost two bits of latch 0000 are 10, so the instruction executed next is instruction_2, in latches 1000, 1001, 1010, and 1011.

1. Latch 1000 holds 0011, so data is copied from latch 0011.
2. Latch 1001 holds 0001, so data is copied to latch 0001.
3. Latch 1010 holds 0001, so only the rightmost bit of the 'to latch,' latch 0001, is copied to.
4. Latch 1011 holds 1100. The rightmost two bits of 1100 are 00 so the data copied from is *not* rotated at all. The leftmost two bits of 1100 are 11, so the next instruction to be executed will be in latches 1100, 1101, 1110, and 1111 (instruction_3).

After instruction_2 is executed, the memory has the following bit values.

	latch address	bit values		
	0000	<u>1100</u>	<------┤	**address of instruction**
	0001	000<u>0</u>	<--┐ ¦	**value of a**
	0010	0001	¦ ¦	**not 0 = 1**
	0011	000*0*	---┘ ¦	**not 1 = 0**
instruction_1	0100	0001	¦	**from address**
	0101	1000	¦	**to address**
	0110	0001	¦	**'to' bits to change**
	0111	1000	¦	**instr.addr.and rot.amount**
instruction_2	1000	0011	¦	**from address**
	1001	0001	¦	**to address**
	1010	0001	¦	**'to' bits to change**
	1011	*1100*	--------┘	**instr.addr.and rot.amount**
instruction_3	1100	0000		**from address**
	1101	0000		**to address**
	1110	0000		**'to' bits to change**
	1111	1100		**instr.addr.and rot.amount**

The italics show from where data was copied. The underlining shows to where data was copied. The arrows show how data was copied.

Latch 0000 now has value <u>11</u>00, so that the next instruction to be executed is instruction_3 in latches <u>11</u>00, <u>11</u>01, <u>11</u>10, and <u>11</u>11.

1. Latch <u>11</u>00 holds 0000, so data will be copied from latch 0000.
2. Latch <u>11</u>01 holds 0000, so data is copied to latch 0000.
3. Latch <u>11</u>10 holds 0000, so <u>*no* data bits</u> are copied (to latch 0000).
4. Latch <u>11</u>11 holds 1100, so 1100 is copied to latch 0000.

This results in the following bit values in memory.

	latch address	Bit values		
	0000	1100	<------	address of instruction
	0001	0000		value of a
	0010	0001		not 0 = 1
	0011	0000		not 1 = 0
instruction_1	0100	0001		from address
	0101	1000		to address
	0110	0001		'to' bits to change
	0111	1000		instr.addr.and rot.amount
instruction_2	1000	0011		from address
	1001	0001		to address
	1010	0001		'to' bits to change
	1011	1100		instr.addr.and rot.amount
instruction_3	1100	0000		from address
	1101	0000		to address
	1110	0000		'to' bits to change
	1111	*1100*	--------	instr.addr.and rot.amount

The underlined bits have been copied to from the italic bits.

Thus, instruction_3 changes nothing (because latch 0000 already held 1100) and leads to instruction_3 being executed again and again.

Ending a program with an instruction like instruction_3 ensures that nothing else will happen after the desired instructions (instruction_1 and instruction_2) are executed. It's just something for the computer to do until we stop the processor and look in memory for the results.

We will now look at some two-instruction programs. The first instruction will do something and the second instruction will do nothing. These short computer programs will show what an instruction (of this simple computer) can do.

Instruction_1 of the following program copies 1111 from latch 0001 to latch 0010. Notice that, because latch 0110 of instrucion_1 holds 1111, all 'to data' bits are copied to. Instruction_2 does nothing over and over.

Before Copy 1111 to 0010 for 1111

	latch address	bit values	
	0000	0100	address of instruction
	0001	1111	from data
	0010	0000	to data
	0011	0000	
instruction_1	0100	0001	from address
	0101	0010	to address
	0110	1111	'to' bits to copy to
	0111	1000	instr.addr.and rot.amount
instruction_2	1000	0000	
	1001	0000	
	1010	0000	
	1011	1000	
	1100	0000	
	1101	0000	
	1110	0000	
	1111	0000	

After instruction_1 is executed, the memory has the following values.

After Copy 1111 to 0010 for 1111

	latch address	bit values		
	0000	1000	<--------- I	address of instruction
	0001	1111	----I	from data
	0010	1111	<---I	to data
	0011	0000		
instruction_1	0100	0001		from address
	0101	0010		to address
	0110	1111		'to' bits to copy to
	0111	1000	----------I	instr.addr.and rot.amount
instruction_2	1000	0000		
	1001	0000		
	1010	0000		
	1011	1000		
	1100	0000		
	1101	0000		
	1110	0000		
	1111	0000		

Instruction_1 of the following program copies 0011 from latch 0001 to latch 0010. Notice that, because latch 0110 of instruction_1 holds 1111, all 'to' bits are copied to.

Before Copy 0011 to 0010 for 0011

	latch address	bit values	
	0000	0100	address of instruction
	0001	0011	from data
	0010	0000	to data
	0011	0000	
instruction_1	0100	0001	from address
	0101	0010	to address
	0110	1111	'to' bits to copy to
	0111	1000	instr.addr.and rot.amount
instruction_2	1000	0000	
	1001	0000	
	1010	0000	
	1011	1000	
	1100	0000	
	1101	0000	
	1110	0000	
	1111	0000	

After instruction_1 is executed, the memory has the following values.

After Copy 0011 to 0010 for 0011

	latch address	bit values		
	0000	1000	<---------	address of instruction
	0001	0011	-----	from data
	0010	0011	<---	to data
	0011	0000		
instruction_1	0100	0001		from address
	0101	0010		to address
	0110	1111		'to' bits to copy to
	0111	1000	----------	instr.addr.and rot.amount
instruction_2	1000	0000		
	1001	0000		
	1010	0000		
	1011	1000		
	1100	0000		
	1101	0000		
	1110	0000		
	1111	0000		

Instruction_1 of the following program copies the rightmost three bits (111) of 1111 from latch 0001 to latch 0010 for 0111. Notice that, because latch 0110 of instruction_1 holds 0111, the rightmost three 'to' bits are copied to.

Before Copy 111 to 0010 for 0111

	latch address	bit values	
	0000	0100	address of instruction
	0001	1111	from data
	0010	0000	to data
	0011	0000	
instruction_1	0100	0001	from address
	0101	0010	to address
	0110	0111	'to' bits to copy to
	0111	1000	instr.addr.and rot.amount
instruction_2	1000	0000	
	1001	0000	
	1010	0000	
	1011	1000	
	1100	0000	
	1101	0000	
	1110	0000	
	1111	0000	

After instruction_1 is executed, the memory has the following values.

After Copy 111 to 0010 for 0111

	latch address	bit values		
	0000	1000	<--------	address of instruction
	0001	1111	----	from data
	0010	0111	<---	to data
	0011	0000		
instruction_1	0100	0001		from address
	0101	0010		to address
	0110	0111		'to' bits to copy to
	0111	1000	----------	instr.addr.and rot.amount
instruction_2	1000	0000		
	1001	0000		
	1010	0000		
	1011	1000		
	1100	0000		
	1101	0000		
	1110	0000		
	1111	0000		

Instruction_1 of the following program copies the rightmost three bits (000) of 0000 from latch 0001 to latch 0010 for 1000. Notice that, because latch 0110 of instruction_1 holds 0111, the rightmost three 'to' bits are copied to.

Before Copy 000 to 0010 for 1000

	latch address	bit values	
	0000	0100	address of instruction
	0001	0000	from data
	0010	1111	to data
	0011	0000	
instruction_1	0100	0001	from address
	0101	0010	to address
	0110	0111	'to' bits to copy to
	0111	1000	instr.addr.and rot.amount
instruction_2	1000	0000	
	1001	0000	
	1010	0000	
	1011	1000	
	1100	0000	
	1101	0000	
	1110	0000	
	1111	0000	

After instruction_1 is executed, the memory has the following values.

After Copy 000 to 0010 for 1000

	latch address	bit values		
	0000	1000	<--------- I	address of instruction
	0001	0000	----I I	from data
	0010	1000	<---I I	to data
	0011	0000	I	
instruction_1	0100	0001	I	from address
	0101	0010	I	to address
	0110	0111	I	'to' bits to copy to
	0111	1000	----------I	instr.addr.and rot.amount
instruction_2	1000	0000		
	1001	0000		
	1010	0000		
	1011	1000		
	1100	0000		
	1101	0000		
	1110	0000		
	1111	0000		

Instruction_1 of the following program rotates the bits (0010) of latch 0001 one space to the left (for 0100) and copies all four rotated bits to latch 0010. Notice that, because latch 0110 of instruction_1 holds 1111, all four bits are copied to. *Also notice that, because latch 0111 has 01 in the rightmost two bits, the from data is rotated one bit to the left.*

Before Rotate 0010 One Bit Left for 0100

	latch address	bit values	
	0000	0100	address of instruction
	0001	0010	from data
	0010	0000	to data
	0011	0000	
instruction_1	0100	0001	from address
	0101	0010	to address
	0110	1111	'to' bits to copy to
	0111	1001	instr.addr.and rot.amount
instruction_2	1000	0000	
	1001	0000	
	1010	0000	
	1011	1000	
	1100	0000	
	1101	0000	
	1110	0000	
	1111	0000	

After instruction_1 is executed, the memory has the following values. Latch 0000 now holds 1001. The right two bits in latch 0000 do *not* affect what instruction is executed next. The left two bits of 1001 (in latch 0000) are 10, so the next instruction to be executed will be instruction_2, in latches 1000, 1001, 1010, and 1011.

After Rotate 0010 One Bit Left for 0100

	latch address	bit values		
	0000	1001	<-------- I	address of instruction
	0001	0010	----I I	from data
	0010	0100	<---I I	to data
	0011	0000	I	
instruction_1	0100	0001	I	from address
	0101	0010	I	to address
	0110	1111	I	'to' bits to copy to
	0111	1001	----------I	instr.addr.and rot.amount
instruction_2	1000	0000		
	1001	0000		
	1010	0000		
	1011	1000		
	1100	0000		
	1101	0000		
	1110	0000		
	1111	0000		

Instruction_1 of the following program rotates the bits (0010) of latch 0001 *two* spaces to the left (for 1000) and copies all four rotated bits to latch 0010. Notice that, because latch 0110 of instruction_1 holds 1111, all four bits are copied to. *Also notice that, because latch 0111 has 10 in the rightmost two bits, the from data is rotated two bits to the left.*

Before Rotate 0010 Two Bits Left for 1000

	latch address	bit values	
	0000	0100	address of instruction
	0001	0010	from data
	0010	0000	to data
	0011	0000	
instruction_1	0100	0001	from address
	0101	0010	to address
	0110	1111	'to' bits to copy to
	0111	1010	instr.addr.and rot.amount
instruction_2	1000	0000	
	1001	0000	
	1010	0000	
	1011	1000	
	1100	0000	
	1101	0000	
	1110	0000	
	1111	0000	

After instruction_1 is executed, the memory has the following values. Again, the right two bits in latch 0000 do *not* affect what instruction is executed next. The left two bits of 1010 (in latch 0000) are 10, so the next instruction to be executed will be instruction_2, in latches 1000, 1001, 1010, and 1011

After Rotate 0010 Two Bits Left for 1000

	latch address	bit values		
	0000	1010	<---------	address of instruction
	0001	0010	----ı	from data
	0010	1000	<---ı	to data
	0011	0000		
instruction_1	0100	0001		from address
	0101	0010		to address
	0110	1111		'to' bits to copy to
	0111	1010	----------ı	instr.addr.and rot.amount
instruction_2	1000	0000		
	1001	0000		
	1010	0000		
	1011	1000		
	1100	0000		
	1101	0000		
	1110	0000		
	1111	0000		

Instruction_1 of the following program rotates the bits (0010) of latch 0001 three spaces to the left (for 0001) and copies all four rotated bits to latch 0010. Notice that, because latch 0110 of instruction_1 holds 1111, all four bits are copied to. *Also notice that, because latch 0111 has 11 in the rightmost two bits, the from data is rotated three bits to the left.* Notice also that rotating three bits to the left is the same as rotating one bit to the right.

Before Rotate 0010 Three Bits Left for 0001

	latch address	bit values	
	0000	0100	address of instruction
	0001	0010	from data
	0010	0000	to data
	0011	0000	
instruction_1	0100	0001	from address
	0101	0010	to address
	0110	1111	'to' bits to copy to
	0111	1011	instr.addr.and rot.amount
instruction_2	1000	0000	
	1001	0000	
	1010	0000	
	1011	1000	
	1100	0000	
	1101	0000	
	1110	0000	
	1111	0000	

After instruction_1 is executed, the memory has the following values.

After Rotate 0010 Three Bits Left for 0001

	latch address	bit values		
	0000	1011	<--------- I	address of instruction
	0001	0010	----I I	from data
	0010	0001	<---I I	to data
	0011	0000	I	
instruction_1	0100	0001	I	from address
	0101	0010	I	to address
	0110	1111	I	'to' bits to copy to
	0111	1011	----------I	instr.addr.and rot.amount
instruction_2	1000	0000		
	1001	0000		
	1010	0000		
	1011	1000		
	1100	0000		
	1101	0000		
	1110	0000		
	1111	0000		

The right two bits of the last word (four bits here) of an instruction indicate how many bits to rotate to the left according to the following table:

bit values	rotate left amount
00	0
01	1
10	2
11	3

If the rightmost bit value is 1, then there is 1 bit of rotation left. If the left bit value is 1, then there is an additional two bits of rotation left.

The following table shows how rotation works with the four bits of a word labeled 'A,' 'B,' 'C,' and 'D.'

rotate bit values	four bits	rotate left amount	rotate right amount
00	ABCD	0	0
01	BCDA	1	3
10	CDAB	2	2
11	DABC	3	1

Notice again that rotating 3 bits left is the same as rotating 1 bit right. Similarly, 1 bit left is 3 bits right and 2 bits left is 2 bits right.

PROCESSOR

Four-Bit Memory

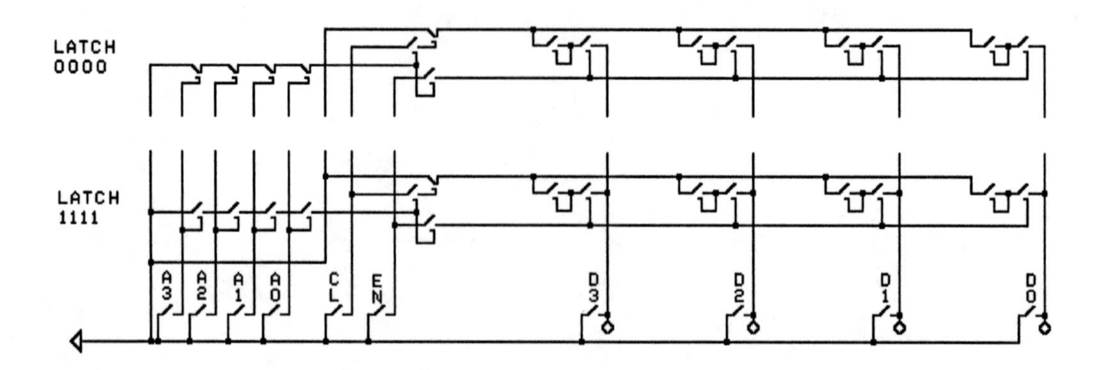

The circuit above shows a memory with four data wires (D3, D2, D1, and D0) and four address wires (A3, A2, A1, and A0). Because there are four address wires, there are sixteen possible latch addresses: 0000, 0001, 0010, 0011, 0100, 0101, 0110, 0111, 1000, 1001, 1010, 1011, 1100, 1101, 1110, and 1111. Only two latches, 0000 and 1111, are shown. The rest are implied by the gap in the circuit diagram.

Two Memories Connected

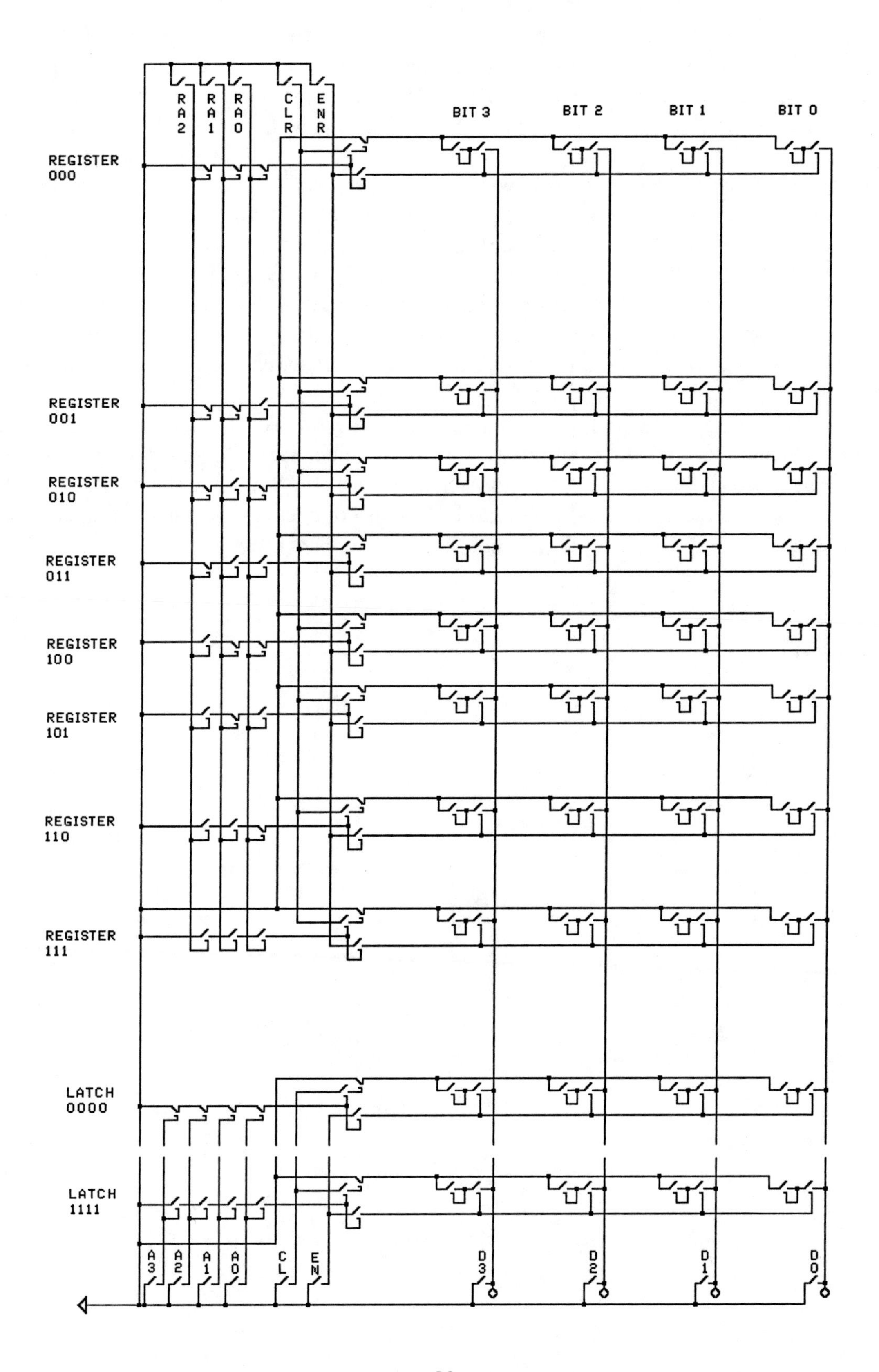

The circuit diagram above shows a memory with four address wires on the bottom connected with a memory with three address wires on the top. Room has been left in the top memory for additional circuitry later. The two memories share data wires D3, D2, D1, and D0. The three-address-wire memory has address wires RA2, RA1, and RA0, clear wire CLR, and enable wire ENR. In the top memory, the latches are called registers and the address wires are called RA2 for Register Address 2, etc. CLR stands for CLear Register. ENR stands for ENable Register.

Because both memories share data wires D3, D2, D1, and D0, data can be copied from a latch of the bottom memory to a register of the top memory or from a register to a latch.

To copy data from a latch to a register, first select the register with register address keys RA2, RA1, and RA0. Second, temporarily press the CLR key to clear the register loops to all 0's. Third, select the latch address with address keys A3, A2, A1, and A0 (while continuing to select the register with RA2, RA1, and RA0). Fourth, temporarily press the enable keys, ENR and EN, to connect the selected register loops and the selected latch loops to the data wires D3, D2, D1, and D0.

Similarly, to copy data from a register to a latch, first select the latch with address keys A3, A2, A1, and A0. Second, temporarily press the clear key, CL (*not* CLR), to clear the *latch*. Third, select the register with register address keys RA2, RA1, and RA0. Fourth, temporarily press the enable keys, ENR and EN. This connects the register and latch loops to the data bus wires, D3, D2, D1, and D0, and, thereby, to each other.

Loops Controlling Lights

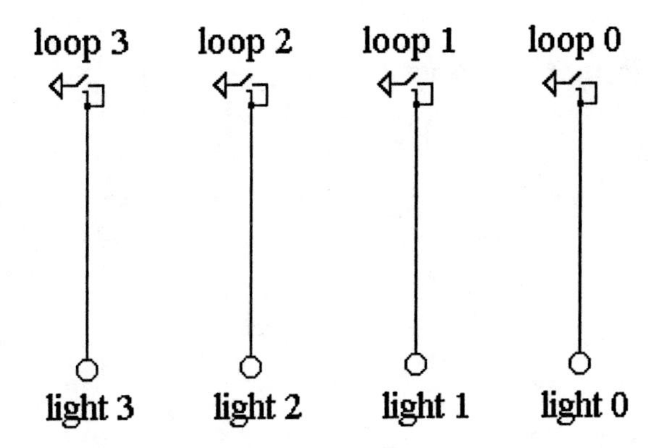

The circuit above shows four loops controlling four lights.

Rotate 1 Circuitry

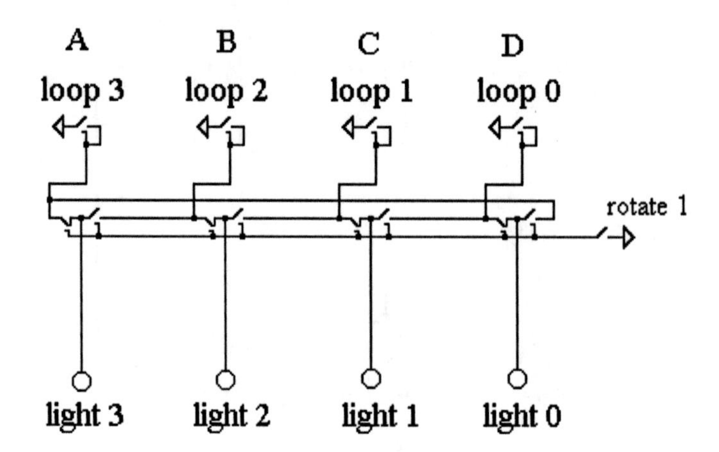

In the circuit above, if the 'rotate 1' key is *not* pressed, then loop3 controls light 3, loop 2 controls light 2, loop 1 controls light 1, and loop 0 controls light 0.

However, if the 'rotate 1' key *is* pressed, then loop 3 controls light 0, loop 2 controls light 3, loop 1 controls light 2, and loop 0 controls light 1. One can say that when the 'rotate 1' key is pressed, then the loop values are rotated one bit to the left. There is *no* bit to the left of bit 3, so bit 3 is rotated to the right end to bit 0.

The following table indicates what pressing the 'rotate 1' key does.

Rotate 1	Light Values				Rotate left amount in bits
	3	2	1	0	
0	A	B	C	D	0
1	B	C	D	A	1

Rotate Two Bits Circuitry

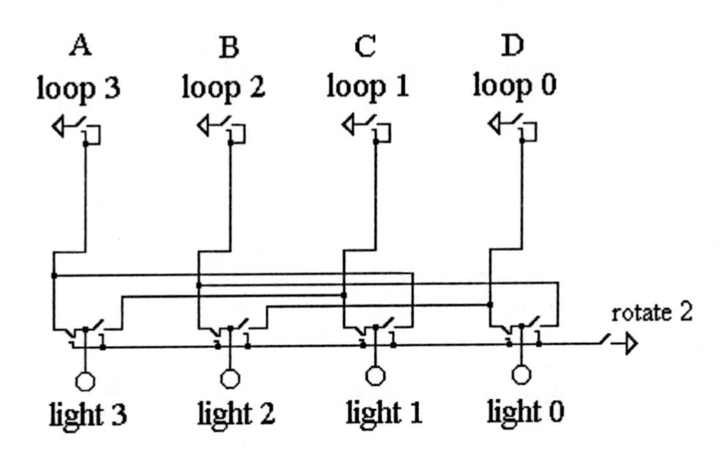

In the circuit above, if the 'rotate 2' key is *not* pressed, then loop 3 controls light 3, loop 2 controls light 2, loop 1 controls light 1, and loop 0 controls light 0.

However, if the 'rotate 2' key *is* pressed, then loop 3 controls light 1, loop 2 controls light 0, loop 1 controls light 3, and loop 0 controls light 2. One can say that when the 'rotate 2' key is pressed, then the loop values are rotated two bits to the left.

The following table indicates what pressing the 'rotate 2' key does.

Rotate 2	Light Values				Rotate left amount in bits
	3	2	1	0	
0	A	B	C	D	0
1	C	D	A	B	2

Rotate Circuitry

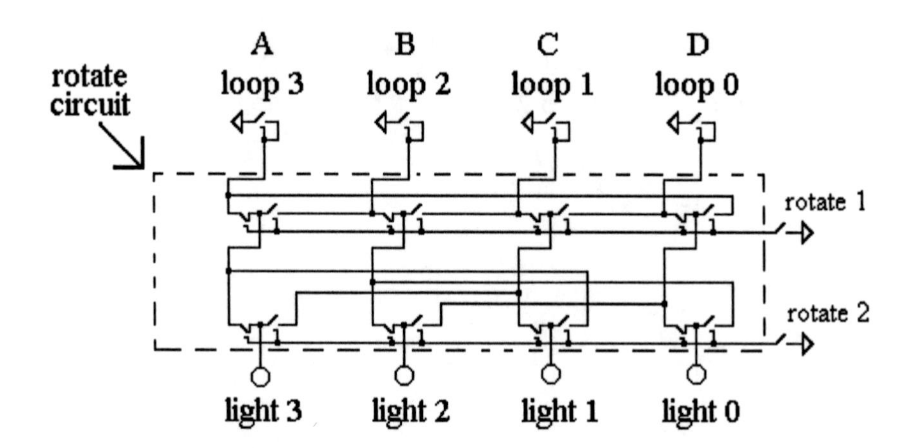

In the circuit above, if *neither* the 'rotate 1' key nor the 'rotate 2' key is pressed, then loop 3 controls light 3, loop 2 controls light 2, loop 1 controls light 1, and loop 0 controls light 0. If rotate 1 is pressed and rotate 2 is *not* pressed, then the loop signals are rotated 1 bit to the left on the way to the lights. If rotate 2 is pressed and rotate 1 is *not* pressed, then the loop values are rotated 2 bits to the left on the way to the lights. Finally, if *both* the rotate 1 key and the rotate 2 key are pressed, then the loop values are rotated three bits to the left. For example, the value in loop 0 is routed to light 3.

The following table indicates what pressing one or both 'rotate' keys does.

Rotate 2	Rotate 1	Light values				Rotate left
		3	2	1	0	amount in bits
0	0	A	B	C	D	0
0	1	B	C	D	A	1
1	0	C	D	A	B	2
1	1	D	A	B	C	3

Mask Circuitry

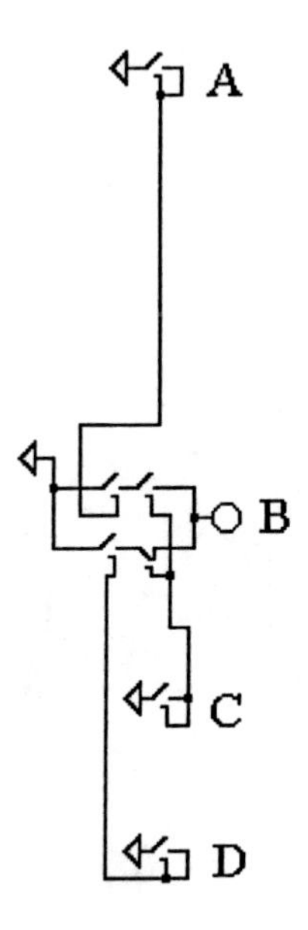

In the circuit above, if loop C has value 0, then light B gets the value in loop D. If loop C has value 1, then light B gets the value in loop A.

Mask Four Bits

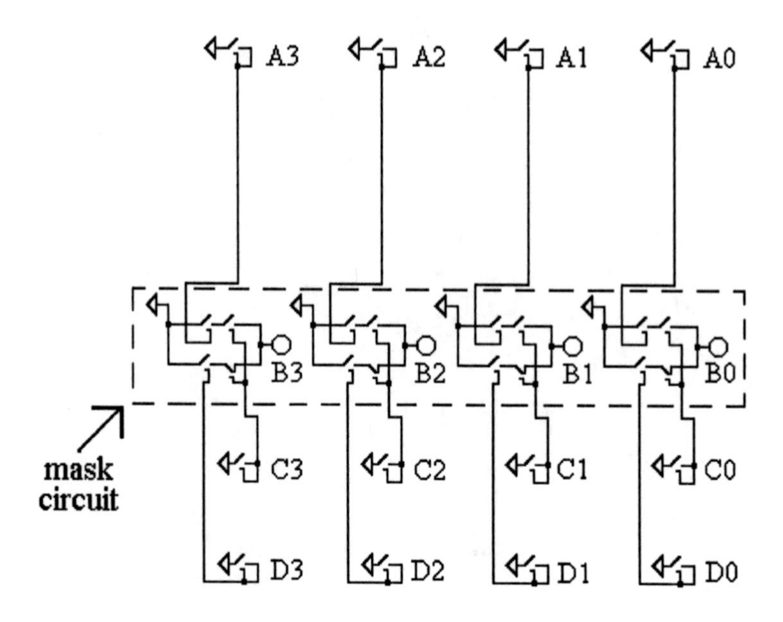

In the left circuit above, C3 controls whether the value of A3 or the value of D3 goes to B3. The other circuits behave similarly.

Rotate and Mask

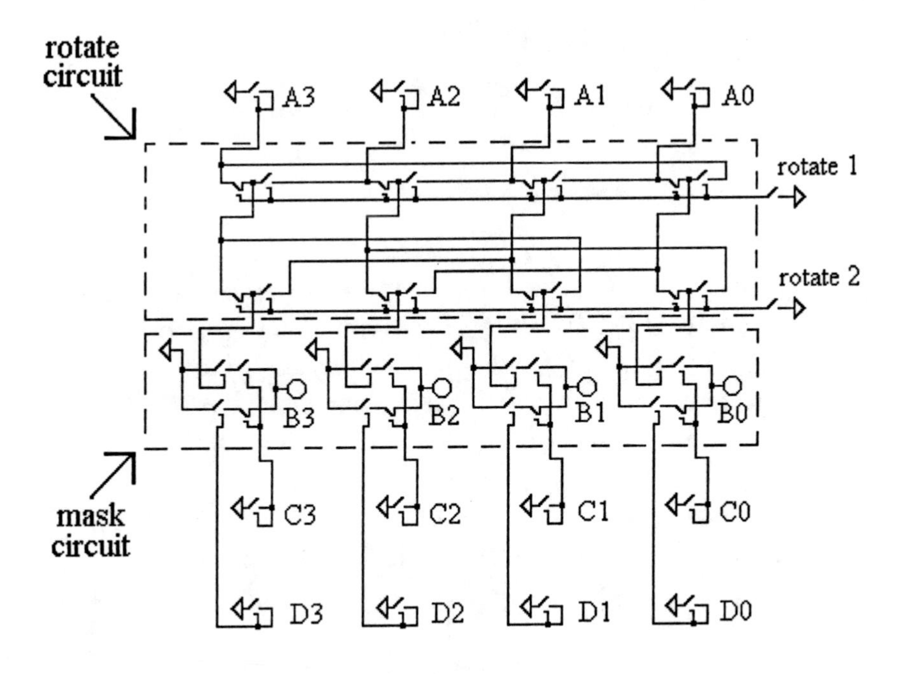

In the circuit above, there is a 'rotate circuit' and a 'mask circuit.'

As an example of the operation of this circuit, consider the case of 'rotate 1' being pressed, 'rotate 2' *not* pressed, C3 = 1, C2 = 1, C1 = 0, and C0 = 0. Then B3 gets A2, B2 gets A1, B1 gets D1, and B0 gets D0. Try to follow the signals in the circuit and see why.

This will be the logic unit of our simple processor. (The logic units of most processors do arithmetic too and so are called arithmetic logic units or, abbreviating, ALU's.)

Delay Circuitry

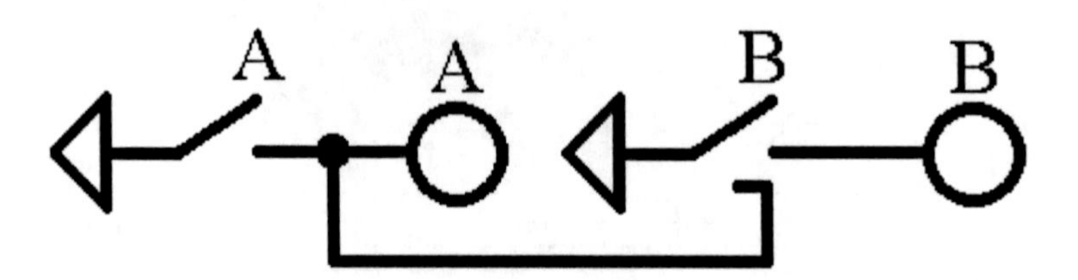

In the circuit above, when key A is pressed, electromagnet B is powered and key B closes. It takes time for B to close after A is pressed. That is, light B comes on about one hundredth of a second after light A. This is indicated by the following 'timing diagram' that shows when the lights come on. Time 0 is 0 seconds. Time 1 is one hundredth of a second (later).

Timing Diagram for Lights

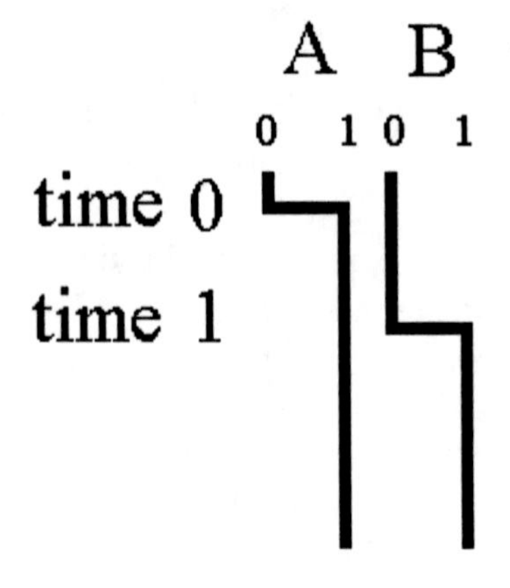

Two Delays

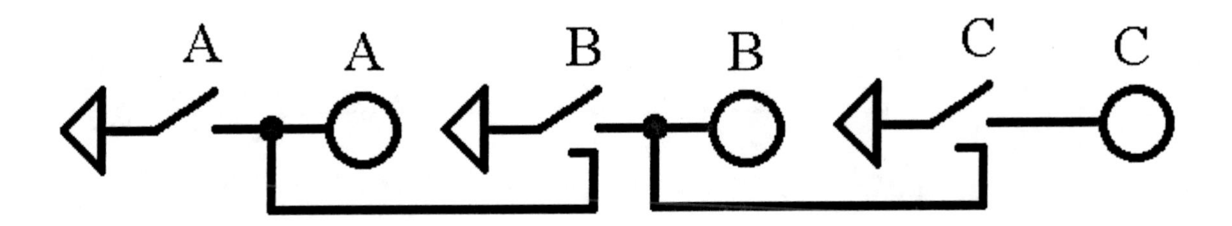

In the circuit above, after key A is pressed it takes one hundredth of a second for key B to close. After key B closes, it takes one hundredth of a second for key C to close. Therefore, after key A is pressed, it takes two hundredths of a second for light C to come on. In the following timing diagram for the circuit above, time 0 is 0 seconds, time 1 is one hundredth of a second, and time 2 is two hundredths of a second.

Timing Diagram for Lights

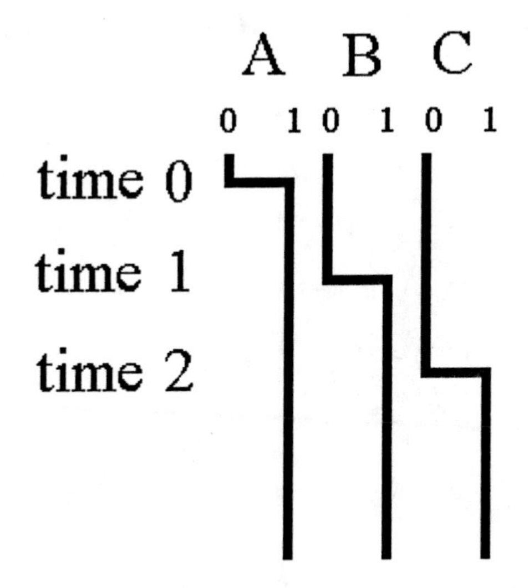

Delay Line

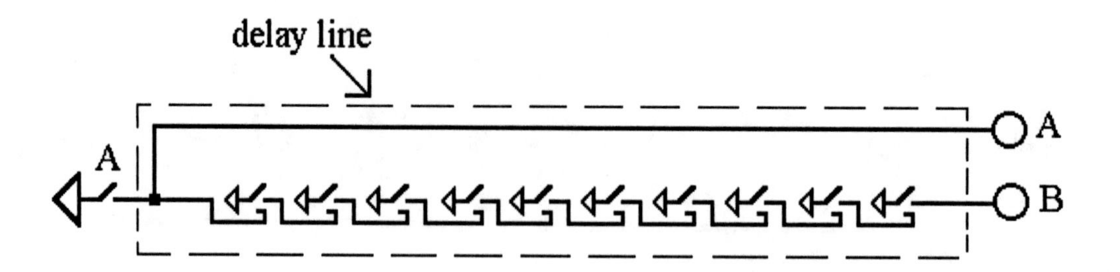

In the circuit (called a delay line) above, light B comes on ten hundredths of a second after light A. Ten hundredths of a second is one tenth of a second, so light B comes on one tenth of a second after key A is pressed (as indicated in the diagram below). (The small amount of time between the time key A is pressed and the time light A comes on is ignored.) Time 0 is 0 seconds and time 1 is one *tenth* of a second in the diagram below.

(When a key closes, it can bounce open and closed a few times. This possible problem will be ignored, except to say that using normally closed relays in a delay line might reduce the problem. This problem does not exist in a delay line made with transistors.)

Timing Diagram for Delay Line

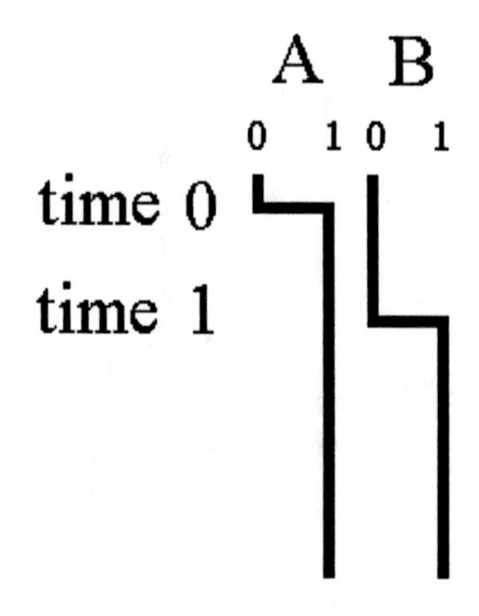

Tapped Delay Line

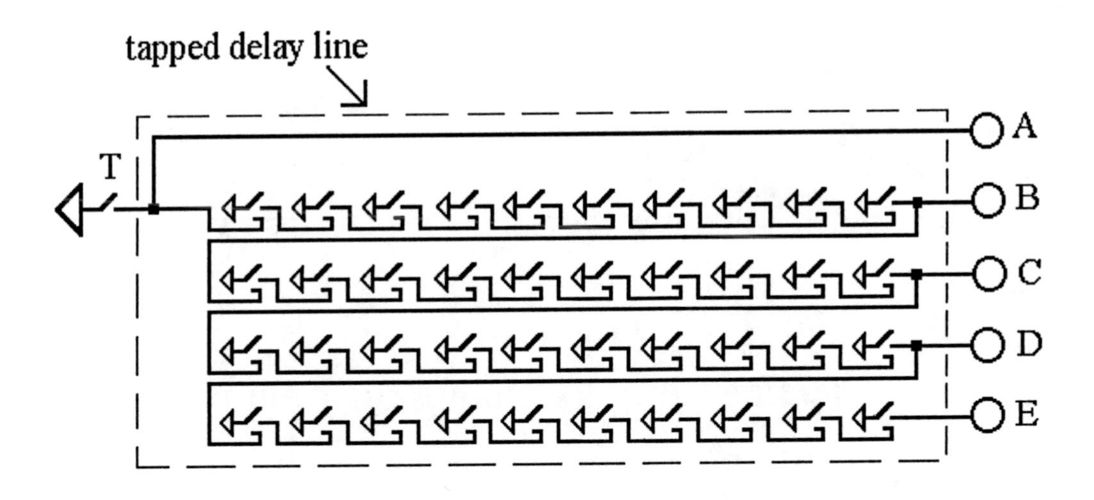

The circuit above is called a tapped delay line. Wires A, B, C, D, and E are called taps. Light B comes on one *tenth* of a second after light A, light C comes on two tenths of a second after light A, light D comes on three tenths of a second after light A, and light E comes on four tenths of a second after light A (as indicated in the timing diagram below).

Timing Diagram for Tapped Delay Line

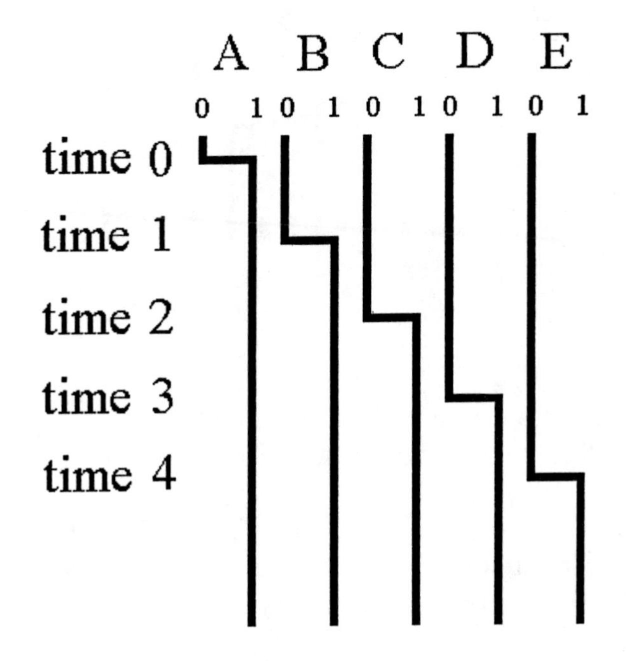

Timing Circuit

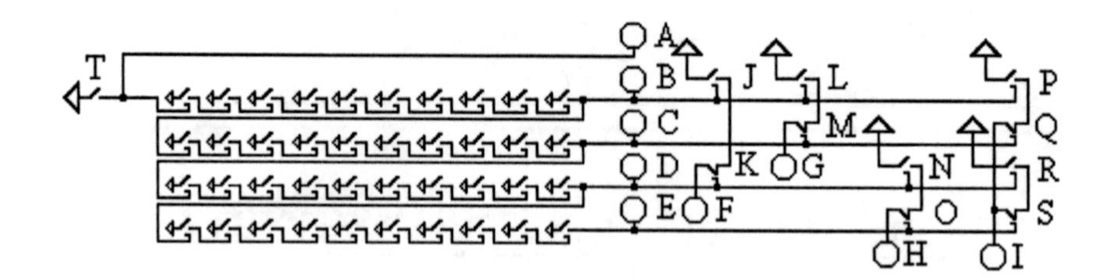

Timing Circuit's Timing Diagram

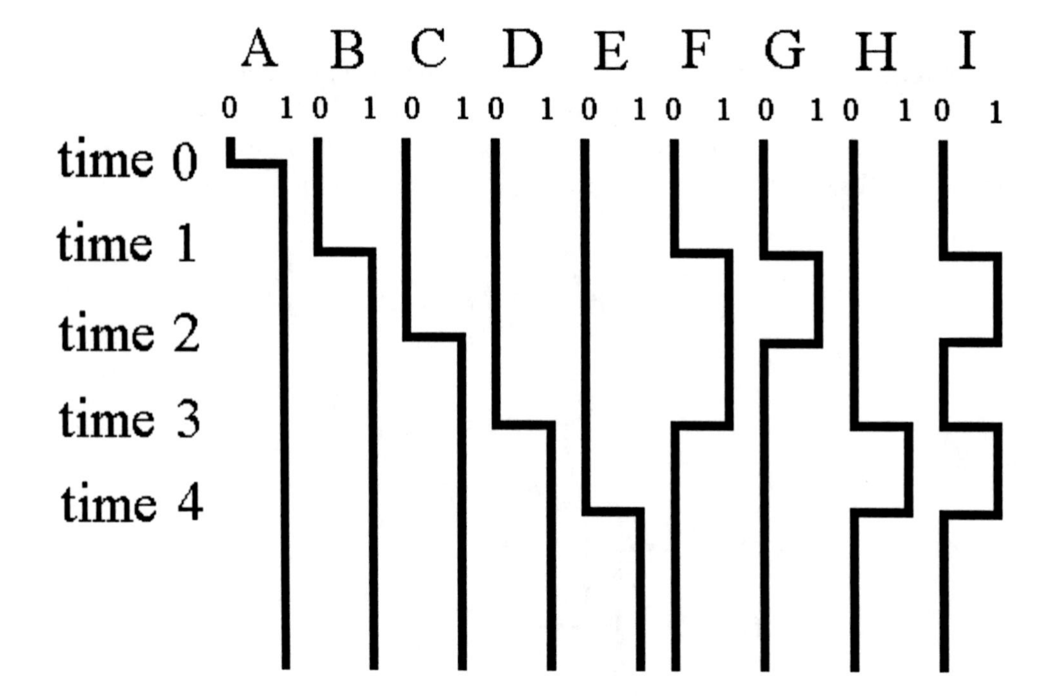

The timing diagram above corresponds to the circuit above. Lights A, B, C, D, and E come on, in order, as before. However, the behaviors of lights F, G, H, and I are more complex.

When light B comes on at time 1, relay J closes. Then electricity can go from the top of the battery (a triangle in the circuit diagram above), through closed relay J and *normally closed relay* K, to light F. Therefore, when light B comes on, light F also comes on. However, when light D comes on at time 3, normally closed relay K *opens* and light F goes out. That is, at time 1, F comes on and, at time 3, F goes out as indicated in the timing diagram.

Similarly, light G comes on when light B comes on, and light G goes out when light C comes on. Similarly, light H turns on when light D comes on, and light H goes off when light E comes on.

The behavior of light I is even more complex. At time 1, light B comes on and relay P closes. Electricity can then go through keys P and Q to light I. At time 2, light C turns on and normally closed relay Q opens, turning light I off. Therefore, light I turns on at time 1, and goes off at time 2. At time 3, light D comes on, relay R closes, and electricity goes from the top of the battery, through key R and the normally closed S key, to light I. At time 4, light E turns on and normally closed relay S opens and light I goes off. Therefore, light I turns on at time 1, off at time 2, on at time 3, and off at time 4.

With Processor Power (PP) Loop

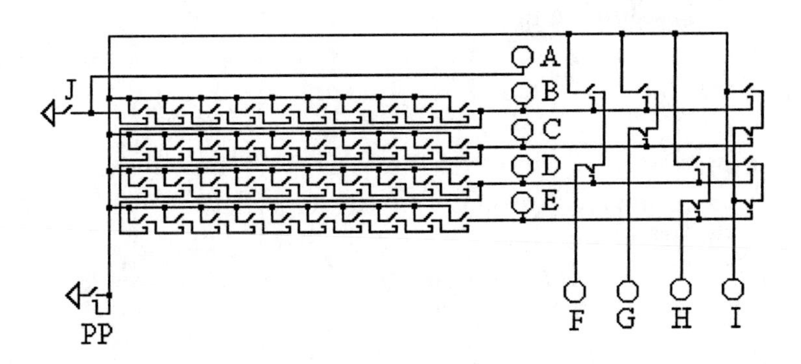

The circuit above is the same as the previous circuit except that all but one connection to power is replaced by a connection to loop 'PP.' ('PP' stands for Processor Power.) After key 'PP' is pressed, key PP stays down and power goes to the circuit. Then, when key J is pressed and held down, output signals F, G, H, and I are generated as indicated in the timing diagram, above. Notice how the right-hand side of the circuit above looks somewhat like the right-hand part of the timing diagram above.

With Feedback Through Normally Closed Key K

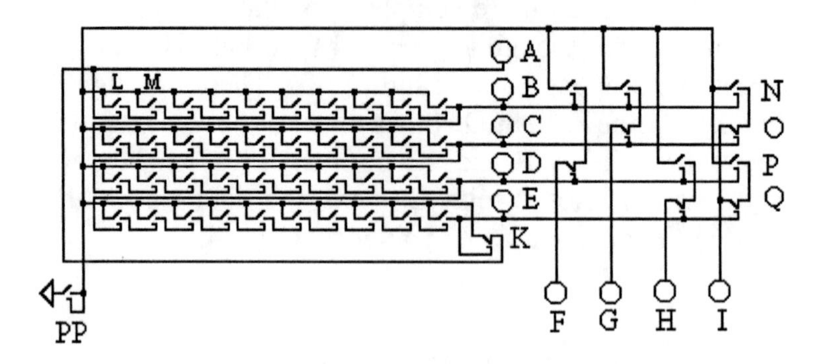

In the circuit above, key J in the upper left has been replaced by the normally closed relay K in the lower center of the circuit. The circuit above generates the timing diagram below when loop key PP in the lower left is pressed at time 0. Loop PP stays down after it is pressed.

When PP is first pressed, electricity can flow from PP, through normally closed relay K to light A and to the electromagnet of relay L in the upper left of the diagram. Relay L then powers relay M. As the relays turn on, one after another, lights B, C, D, and then E turn on. When light E turns on, the *normally closed relay* K opens, light A goes out, and relay L opens. One hundredth of a second after relay L opens, relay M opens because electricity is no longer getting to the electromagnet of relay M. The relays in the delay line then open one after another and lights B, C, D, and E go off one after another. When light E goes off, *no* power gets to the electromagnet of normally closed relay K and relay K closes. When relay K closes, electricity can get to light A and then lights B, C, D, and E turn on.

Thus, A, B, C, D, and E turn on one after another. Then A, B, C, D, and E go off one after another. Then A, B, C, D, and E turn on one after another. Then A, B, C, D, and E go off one after another. This pattern repeats as long as loop key PP stays down.

Light F is on only when light B is on and light D is off. Similarly, light G is on only when light B is on and light C is off. Also similarly, light H is on only when light D is on and light E is off.

When light B is on and light C is off, relays N and O are closed and light I is on. Similarly, when light D is on and light E is off, relays P and Q are closed and light I is on. Therefore, light I is only on when light B is on and light C is off *and* when light D is on and light E is off.

The circuit above is called a clock. It generates signals F, G, H, and I over and over again as indicated in the timing diagram below.

Timing Diagram with Feedback

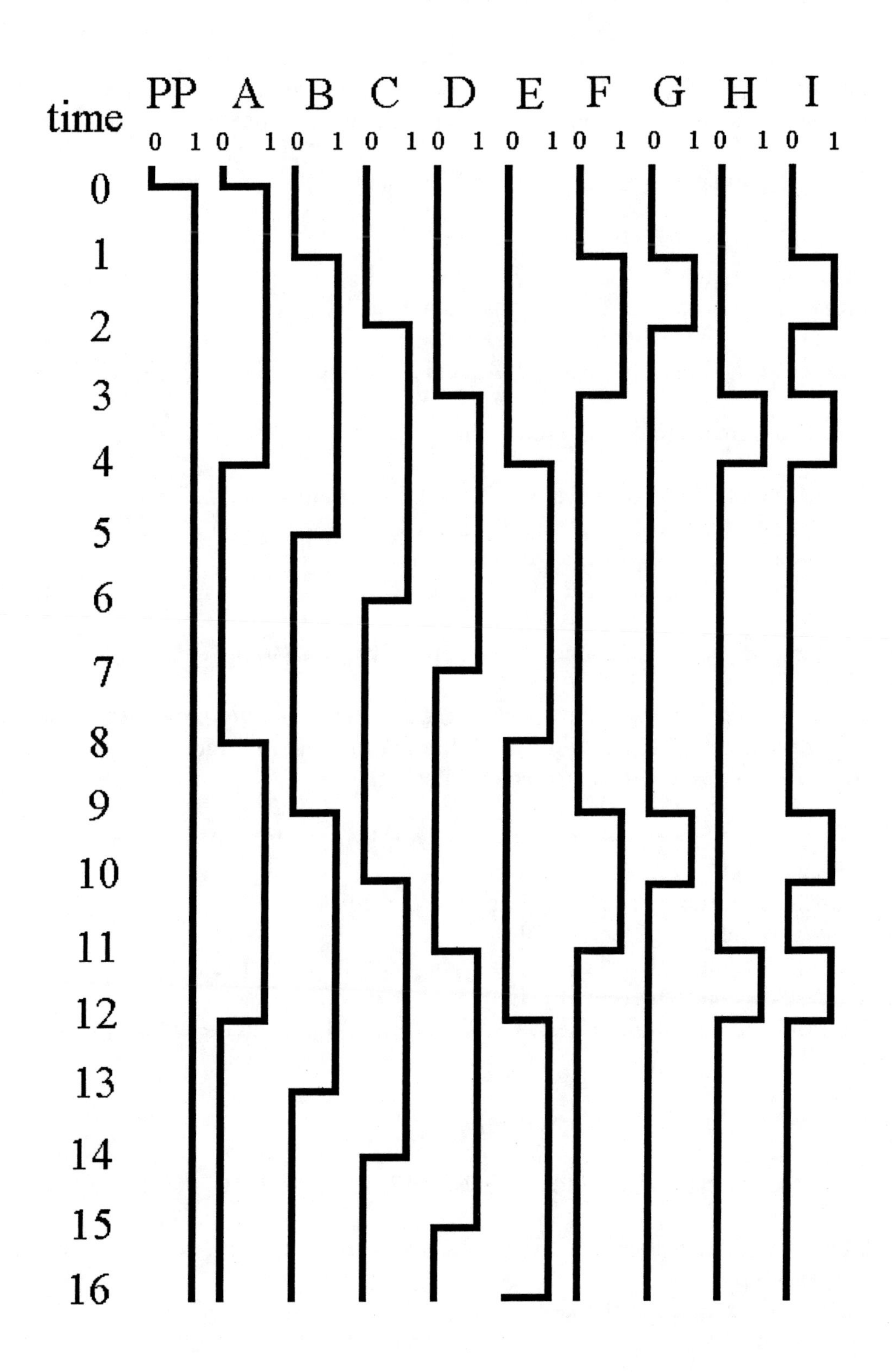

The circuit below shows two latches of memory at the bottom, latch 0000 and latch 1111. The other fourteen latches are not shown. It also shows a processor above the memory. The processor is mainly made of latches. Latches in a processor are called registers. Register 001 is not a latch, however, because it doesn't have loops.

Writing to a latch will *not* clear *any* bits that were previously 1, so *always* clear a latch before writing data to it. Therefore, to write to a latch, do the following.

1. Press the correct address keys (A3, A2, A1, and A0) and data keys (D3, D2, D1, and D0).
2. Press the clear key, CL, to clear the latch.
3. Release the clear key, CL.
4. Press the enable key, EN, to copy data to the latch.
5. Release the enable key, EN.
6. Release the address keys, A3, A2, A1 and A0, and the data keys, D3, D2, D1 and D0.

To *read* data from a latch, do the following.

1. Press the correct address keys, A3, A2, A1 and A0, to select the latch to read.
2. Press the enable key, EN, to send the latch's values to the lights, D3, D2, D1, and D0.
3. Release the enable key EN.
4. Release the address keys, A3, A2, A1 and A0.

Notice that reading a latch connects the loops of the latch to the data bus wires, D3, D2, D1, and D0.

Similarly, writing to a register will *not* clear any bits that were previously 1, so *always* clear a register before writing data to it. Therefore, to write to a register (except 'register' 001, which has no loops to write to), do the following.

1. Press the correct register address keys (RA2, RA1, and RA0) and data keys (D3, D2, D1, and D0).
2. Press the clear register key, CLR, to clear the register.
3. Release the clear register key, CLR.
4. Press the enable register key, ENR, to copy data to the register.
5. Release the enable register key, ENR.
6. Release the register address keys RA2, RA1 and RA0, and the data keys, D3, D2, D1 and D0.

To *read* register data, just do the following.

1. Press the correct register address keys, RA2, RA1 and RA0, to select the register to read.
2. Press the enable register key, ENR, to copy the register's values to the lights D3, D2, D1, and D0.
3. Release the enable register key, ENR.
4. Release the register address keys, RA2, RA1 and RA0.

Notice that reading a register connects the loops of the register to the data bus wires, D3, D2, D1, and D0.

Do the following to copy data from a latch to a register (except for register 001).

1. Select the register with RA2, RA1, and RA0.
2. Temporarily press CLR to clear the register.
3. Select the latch in memory with A3, A2, A1, and A0.
4. Temporarily press ENR and EN to connect the register loops and latch loops to the data bus wires, D3, D2, D1, and D0, and so to each other.
5. Release all keys.

Do the following to copy data from a register to a latch.

1. Select the latch with A3, A2, A1, and A0.
2. Temporarily press CL to clear the latch.
3. Select the register with RA2, RA1, and RA0.
4. Temporarily press ENR and EN to connect the register loops and latch loops to the data bus wires, D3, D2, D1, and D0, and so to each other.
5. Release all keys.

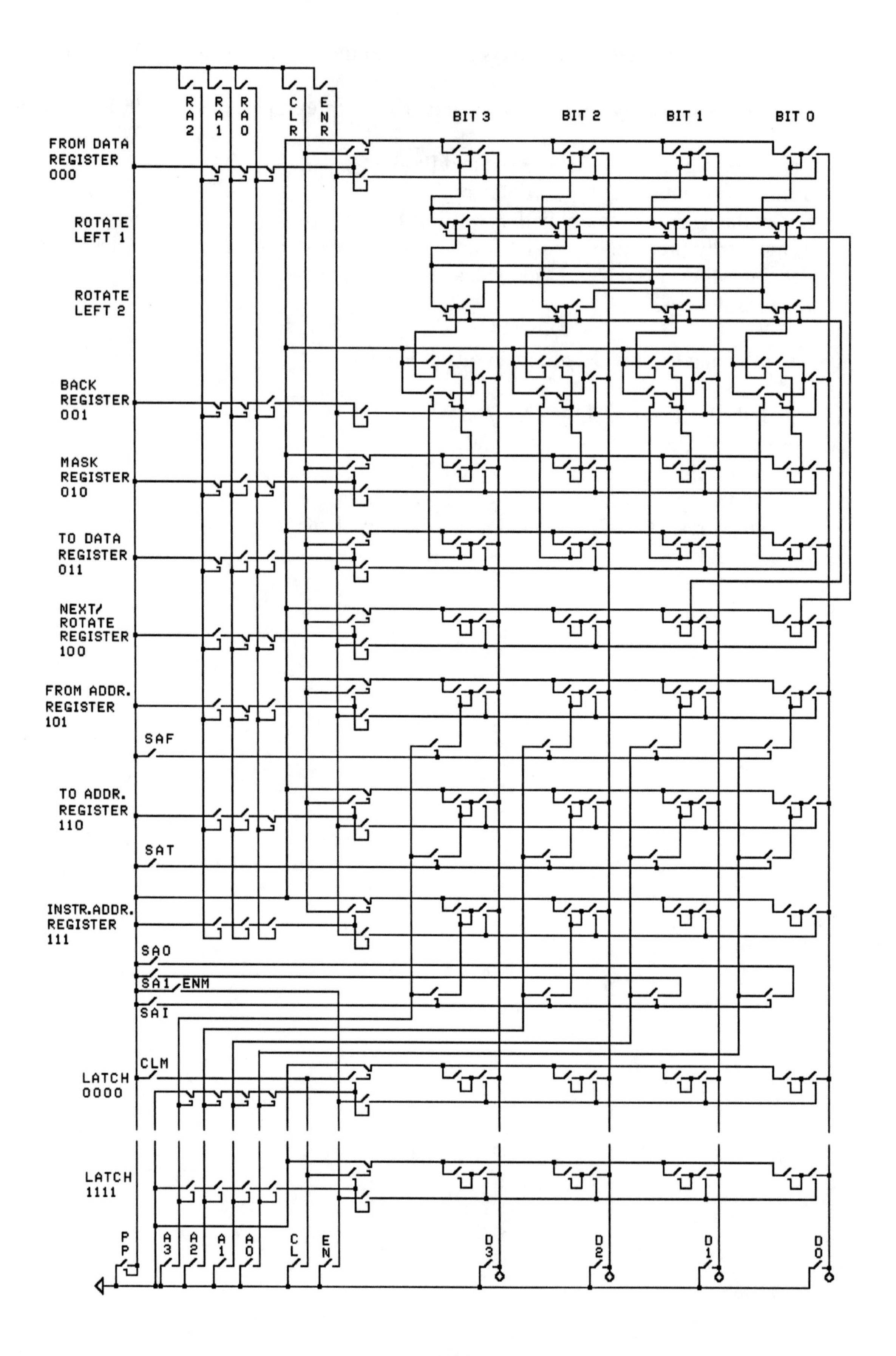

R A 2
R A 1
R A 0
C L R
E N R
BIT 3
BIT 2
BIT 1
BIT 0
FROM DATA REGISTER 000
ROTATE LEFT 1
ROTATE LEFT 2
BACK REGISTER 001
MASK REGISTER 010
TO DATA REGISTER 011
NEXT/ ROTATE REGISTER 100
FROM ADDR. REGISTER 101
SAF
TO ADDR. REGISTER 110
SAT
INSTR.ADDR. REGISTER 111
SA0
SA1
ENM
SAI
CLM
LATCH 0000
LATCH 1111
P P
A 3
A 2
A 1
A 0
C L
E N
D 3
D 2
D 1
D 0

An instruction is executed in nine steps. Step 1 copies the address of the instruction from latch 0000 of the memory to the processor. Steps 2, 3, 4, and 5 copy the four words of the instruction to the processor. Step 6 copies the 'from data' to the processor. Step 7 copies the 'to data' to the processor. Step 8 copies the result from the processor to the memory. Step 9 copies the address of the next instruction to be executed from the processor to latch 0000 of the memory.

The timing diagram that follows, as well as the explanation that follows, tells the order in which to press the keys to execute an instruction with the circuit in the diagram above. *Look at the timing diagram below and the circuit above as you read about each step.* The following nine steps are tedious, but try to get through them or, at least, study step 1 and read through the rest.

1. The first step in executing an instruction is to copy the value in latch 0000 to register 111. Latch 0000 holds the address of the instruction to be executed. To copy the contents of latch 0000 to register 111, the following is done.

First, RA2 is pressed (gets set to 1), RA1 gets 1, and RA0 gets 1. This selects register 111, the instruction address register. SAF, SAT, and SAI are each set to 0 (not pressed), which selects latch 0000. (This will become clear later.) Then, CLR (CLear Register) is temporarily pressed to clear register 111. Then ENM (ENable Memory) and ENR (ENable Register) are temporarily pressed. When ENM and ENR are pressed, the loops of both latch 0000 and register 111 are connected to the data bus. Therefore, the values in latch 0000 can flow to the loops of (just cleared) register 111 causing register 111 to have the same values as latch 0000. Register 111 then (also) holds the address of the instruction to be executed next.

2. The second step in executing an instruction is to copy the first word (four bits) of the instruction to register 101, the 'from address register,' because the first four bits (word) of an instruction are the address from which the data will be copied. RA0 and RA2 are pressed and RA1 is released to select register 101. Key SAI, for Select Address of Instruction, is pressed, routing the left two bits of register 111 to the address wires, A3 and A2. Pressing key SAI also routes the values of SA1 (Select Address bit 1) and SA0 (Select Address bit 0) to A1 and A0 of the memory. SA1 and SA0 are not pressed, so A1 and A0 get 0. Next, CLR is temporarily pressed, thereby clearing register 101. Next, ENR and ENM are temporarily pressed, connecting to data bus register 101 and the first latch (whose address ends in 00) of the instruction to be executed. Thus the first word of the instruction is copied to latch 101.
3. Third, the second word of the instruction is copied to 'to address register 110.' This is done by pressing RA2, pressing RA1, and *not* pressing RA0. Also SAI is pressed, SA1 (Select Address bit 1) is *not* pressed, and SA0 is pressed. Then CLR is temporarily pressed to clear register 110. Then ENM and ENR are temporarily pressed to copy the contents of the second word (4 bits) of the instruction to register 110.
4. Fourth, the third word of the instruction is copied to 'mask register 010.' This is done by *not* pressing RA2, pressing RA1, and *not* pressing RA0. Also SAI (Select Address of Instruction) is pressed, SA1 is pressed, and SA0 is *not* pressed. Then CLR is temporarily pressed to clear register 010. Then ENM and ENR are temporarily

pressed to copy the contents of the third word (4 bits) of the instruction to register 010.

5. Fifth, the fourth word of the instruction is copied to 'next/rotate register 100.' This is done by pressing RA2, *not* pressing RA1, and *not* pressing RA0. Also SAI is pressed, SA1 is pressed, and SA0 is pressed. Then CLR is temporarily pressed to clear register 100. Then ENM and ENR are temporarily pressed to copy the contents of the fourth word (4 bits) of the instruction to register 100.
6. Sixth, RA2, RA1, and RA0 are *not* pressed to select register 000, the 'from data register.' SAF (Select Address of From data) is pressed to route the address in the 'from address register 101' to the memory's address wires A3, A2, A1, and A0. CLR is then temporarily pressed to clear register 000. Next, ENM and ENR are temporarily pressed to copy the contents of memory *pointed to by* 'from address register 000' to 'from data register 000.'
7. Seventh, RA2 is *not* pressed, RA1 is pressed, and RA0 is pressed to select 'to data register 011.' SAT (Select Address of To data) is pressed to route the address in 'to address register 110' to the memory's address wires A3, A2, A1, and A0. CLR is then temporarily pressed to clear register 011. Next, ENM and ENR are temporarily pressed to copy the contents of memory pointed to by 'to *address* register 110' to 'to *data* register 011.'
8. Eighth, RA2 is *not* pressed, RA1 is *not* pressed, and RA0 is pressed to select 'back data register 001.' SAT is pressed to route the address in 'to address register 110' to the memory's address wires, A3, A2, A1, and A0. Next, CLM (CLear Memory, *not* CLR, CLear Register) is temporarily pressed to clear the latch *in memory* pointed to by 'to address register 110.' ENR and ENM are then temporarily pressed to copy some rotated bits of 'from data register 000' and *not* rotated bits of 'to data register 011' to the address in memory pointed to by 'to address register 110.' If the rightmost bit of 'next/rotate register 100' is 1, then the from data is rotated 1 bit left. If the second-to-rightmost bit of 'next/rotate register 100' is 1, then the from data is rotated an additional 2 bits left. If a bit of 'mask register 010' is 0, then the corresponding bit of 'to data register 011' is copied back to memory. However, if a bit of 'mask register 010' is 1, then the corresponding rotated bit of 'from data register 000' is copied back to memory. Notice that because CLM was pressed instead of CLR, a latch of memory was cleared and copied to instead of a register.
9. Ninth, RA2 is pressed, RA1 is *not* pressed, and RA0 is *not* pressed to select 'next/rotate register 100.' SAI, SAF, and SAT are *not* pressed, so *no* address goes to the memory, so latch 0000 is selected. Next, CLM (*not CLR*) is temporarily pressed to clear latch 0000 in memory. ENR and ENM are then temporarily pressed to copy the data in 'next/rotate register 100' to latch 0000 in memory. This prepares for the next instruction. Notice, again, that because CLM was pressed instead of CLR, a latch of memory was cleared and copied to instead of a register.

Timing Diagram for Instruction

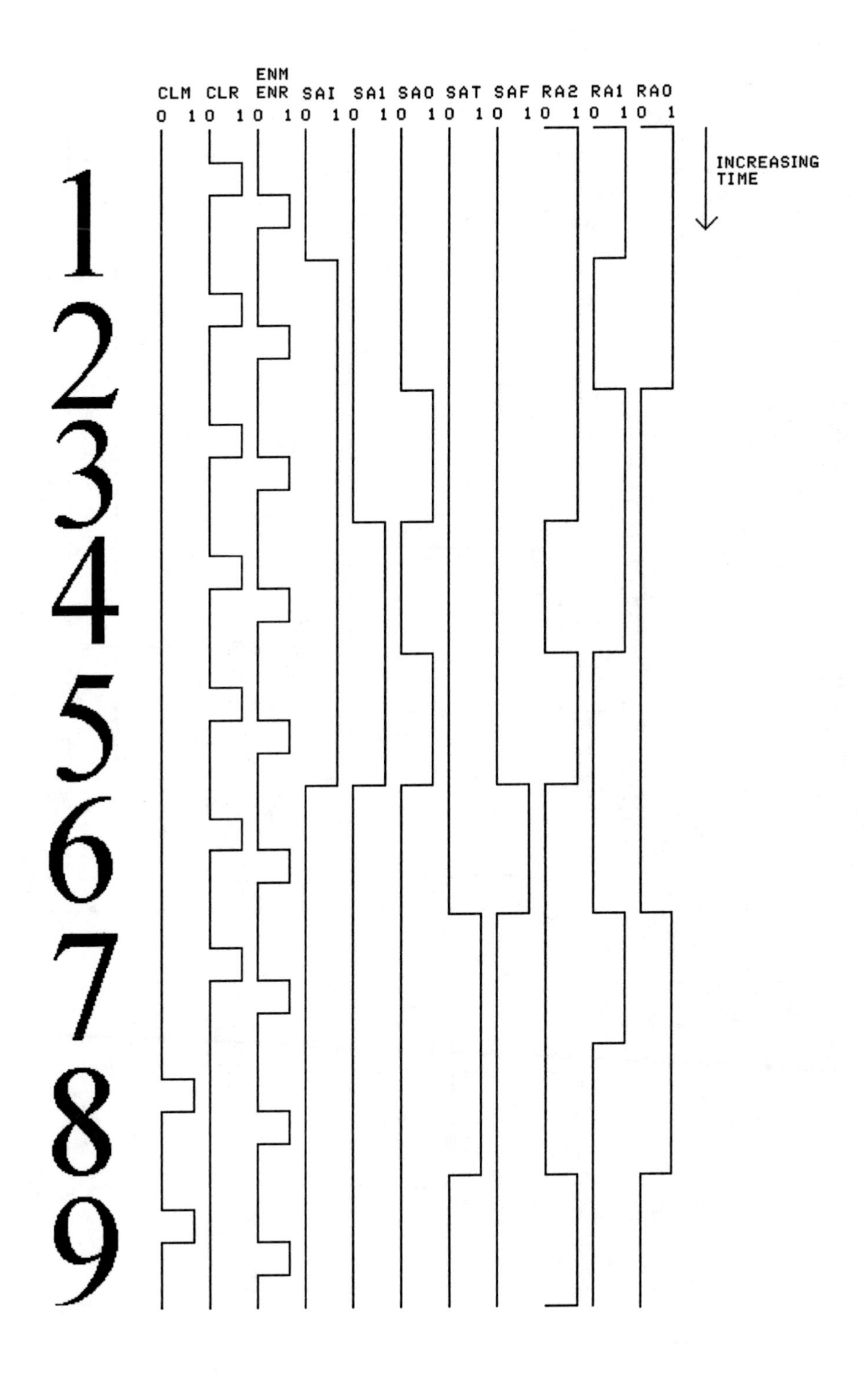

Clock Circuit for Processor

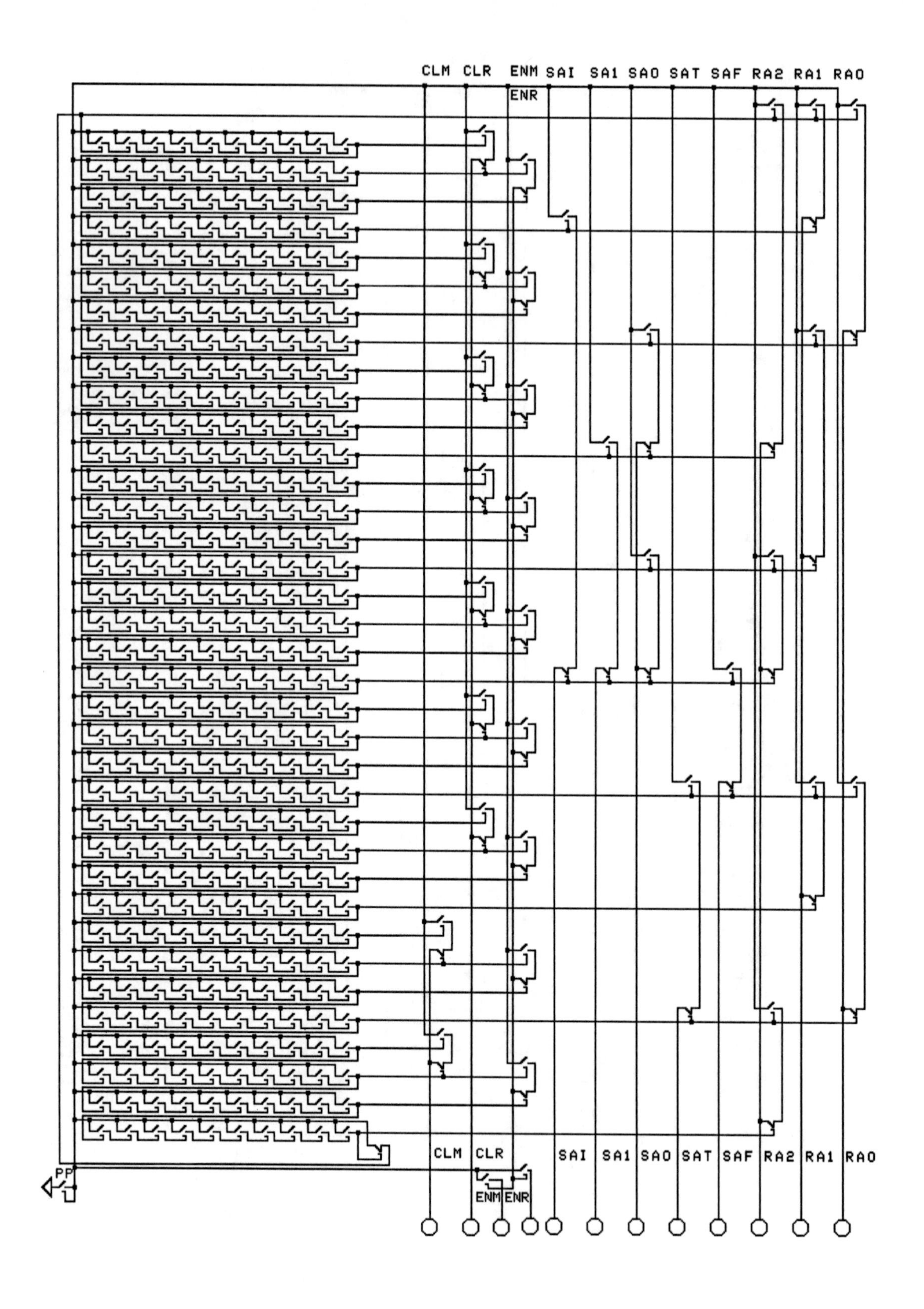

The preceding clock circuit shows a circuit that repeatedly generates the timing diagram signals. The outputs of this circuit can be connected to the processor to make the processor repeatedly execute instructions as indicated in the diagram of the complete (though simple) computer in the diagram below.

The operation of a clock has already been explained.

It takes about 361 hundredths of a second for the timing diagram to be generated. Then all outputs of the right hand side are 0 for another about 361 hundredths of a second. Then the timing diagram is generated again, etc. Therefore, it takes this computer about 722 hundredths of a second to execute each instruction! This is one main reason that transistors are now used. Transistors are millions of times faster. The other reason is that a relay costs as much as millions of interconnected transistors. However, a transistor-based computer works in the same way as a relay-based computer. The cheapness of transistors allows much more memory. It also allows extra things to be added to the processor like more registers and extra circuits to do certain common things, like multiply two numbers together, more quickly.

The whole computer is illustrated on the following two pages. The clock, processor and memory are shown and interconnected. The processor includes the rotate and mask circuitry.

To use the computer below, first enter the program and data into the memory with the keys at the bottom of the circuit: A3, A2, A1, A0, CL, EN, D3, D2, D1, and D0. Then, press PP at the bottom of the circuit to make the computer run. Wait until the program is finished and lift up key PP. Then use the keys at the bottom of the circuit to read the results from memory.

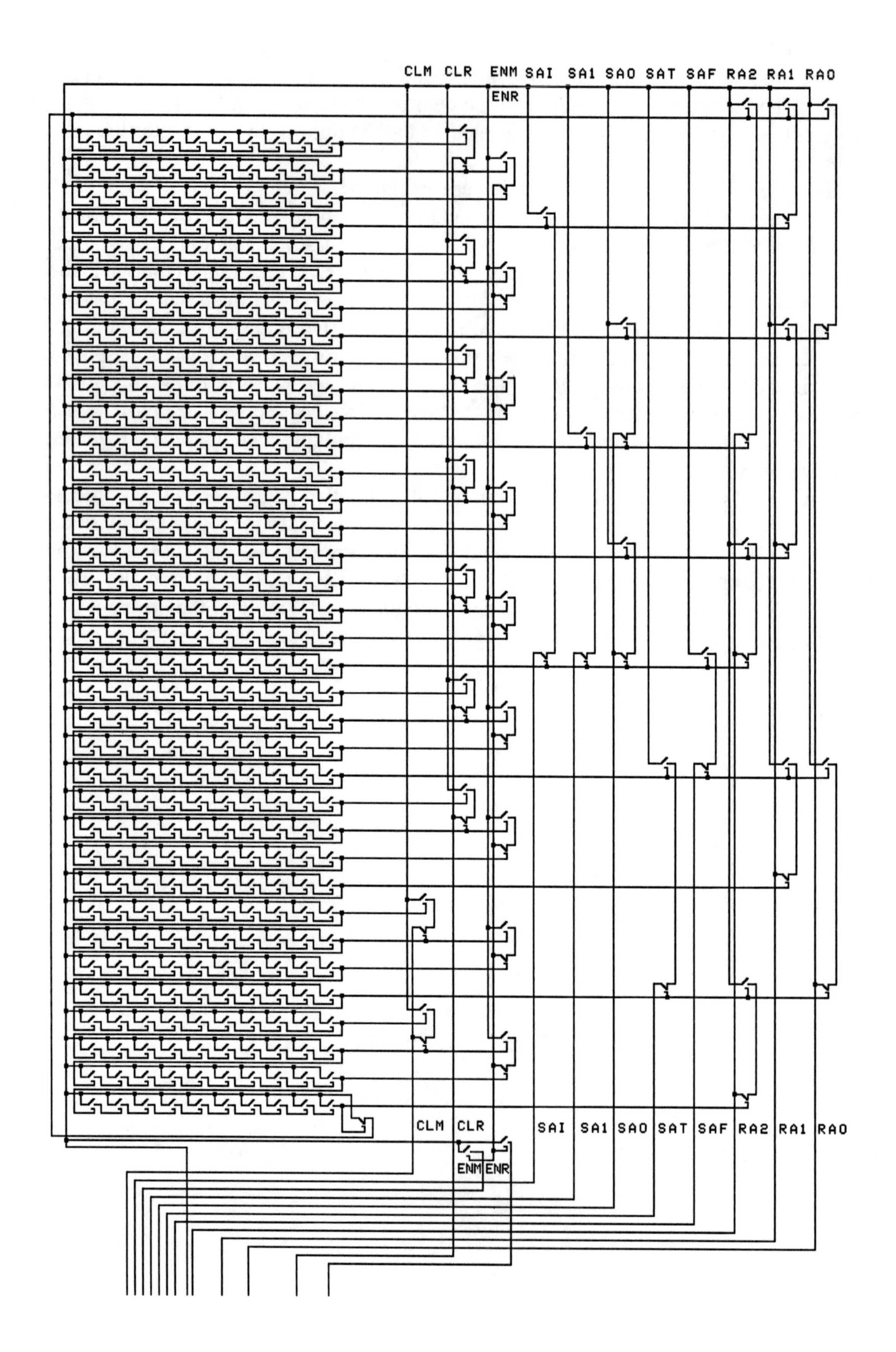
CLM
CLR
ENM
ENR
SAI
SA1
SA0
SAT
SAF
RA2
RA1
RA0
CLM
CLR
SAI
SA1
SA0
SAT
SAF
RA2
RA1
RA0
ENM
ENR

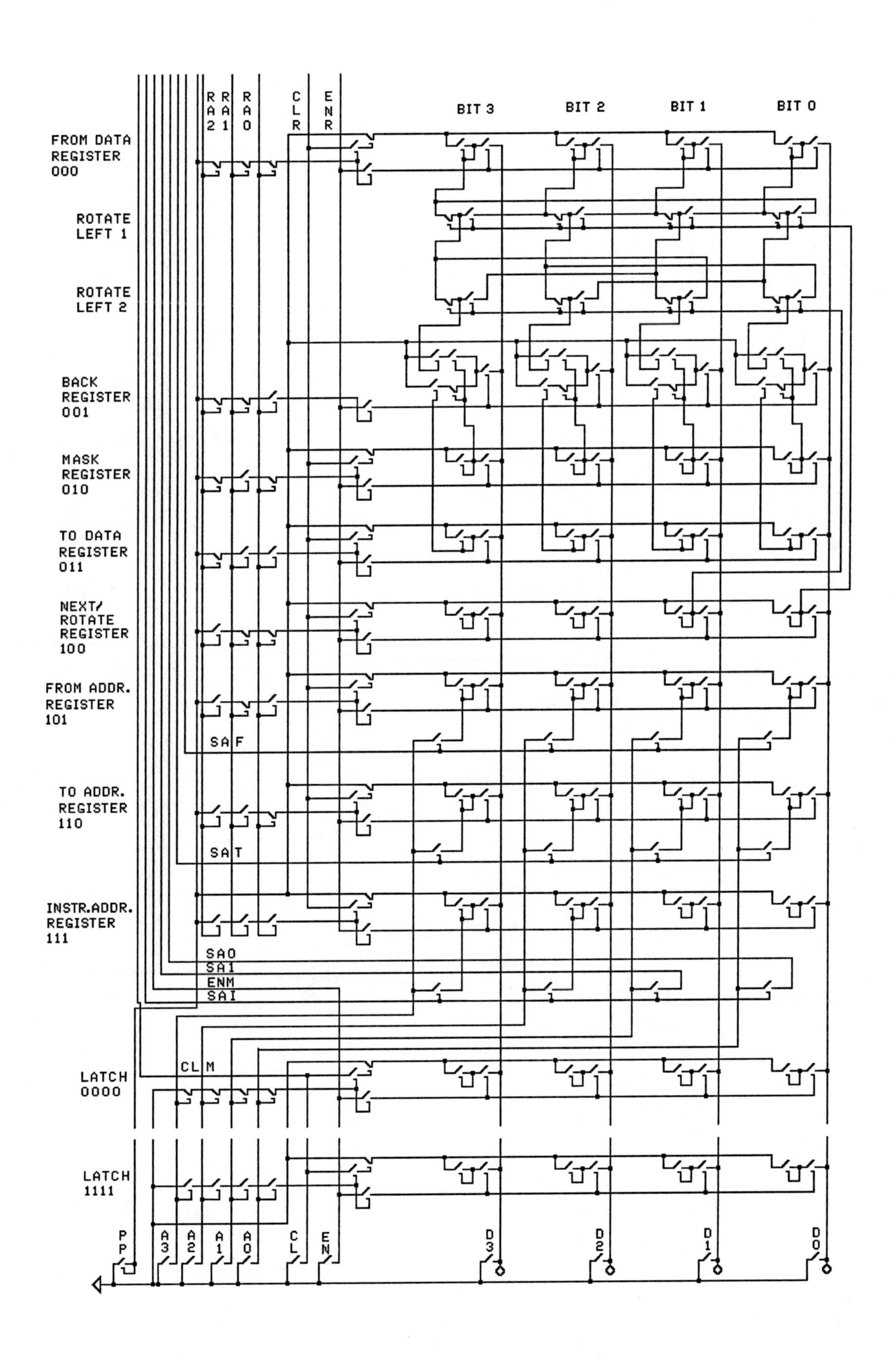
RA2
RA1
RA0
CLR
ENR
BIT 3
BIT 2
BIT 1
BIT 0
FROM DATA REGISTER 000
ROTATE LEFT 1
ROTATE LEFT 2
BACK REGISTER 001
MASK REGISTER 010
TO DATA REGISTER 011
NEXT/ ROTATE REGISTER 100
FROM ADDR. REGISTER 101
SA F
TO ADDR. REGISTER 110
SA T
INSTR.ADDR. REGISTER 111
SA0
SA1
ENM
SAI
CL M
LATCH 0000
LATCH 1111
PP
A3
A2
A1
A0
CL
EN
D3
D2
D1
D0

PROGRAMMING

We will now consider how to program a 16-bit rather than 4-bit computer.

A 16-bit computer has 16 bits in each 'word' and 65536 words of memory. This is because there are 65536 possible 16-bit addresses: 0000000000000000, 0000000000000001, 0000000000000010, 0000000000000011, 0000000000000100, etc.

The instruction still consists of four words, but now each instruction is 16 bits long. An example instruction is:

Label	**address**	**data**	**comment**
instr_1	**0000000000000100**	**0000000000000001**	**from address**
	0000000000000101	**0000000000000010**	**to address**
	0000000000000110	**0000000011111111**	**'to' bits to copy to**
	0000000000000111	**000000000010*0000***	**instr.addr.and *rot.amount***

The labels and comments are *not* part of the program. The addresses just show where the data is stored. *The data is the program.* An instruction written with instructions that are just 1's and 0's is a machine language instruction. Machine language instructions are called 'machine code.'

'Instr.addr.and *rot.amount*' is short for 'next instruction's address and *rotate amount*.'

Notice that there are now four (*italic*) rotate bits (*0000*). They cause the copied-from data to be rotated according to the following table.

16-Bit Rotate Table

rotate bit values	**16 bits**	**rotate left amount**	**rotate right amount**
0000	**ABCDEFGHIJKLMNOP**	**0**	**0**
0001	**BCDEFGHIJKLMNOPA**	**1**	**15**
0010	**CDEFGHIJKLMNOPAB**	**2**	**14**
0011	**DEFGHIJKLMNOPABC**	**3**	**13**
0100	**EFGHIJKLMNOPABCD**	**4**	**12**
0101	**FGHIJKLMNOPABCDE**	**5**	**11**
0110	**GHIJKLMNOPABCDEF**	**6**	**10**
0111	**HIJKLMNOPABCDEFG**	**7**	**9**
1000	**IJKLMNOPABCDEFGH**	**8**	**8**
1001	**JKLMNOPABCDEFGHI**	**9**	**7**
1010	**KLMNOPABCDEFGHIJ**	**10**	**6**
1011	**LMNOPABCDEFGHIJK**	**11**	**5**
1100	**MNOPABCDEFGHIJKL**	**12**	**4**
1101	**NOPABCDEFGHIJKLM**	**13**	**3**
1110	**OPABCDEFGHIJKLMN**	**14**	**2**
1111	**PABCDEFGHIJKLMNO**	**15**	**1**

The bits to the left of the rotate bits in latch 0000000000000111 are 000000000010 and indicate that the next instruction will be in latches 0000000000001000, 0000000000001001, 0000000000001010, and 0000000000001011.

Latch 0000000000000100 holds 0000000000000001, so data is copied *from* latch 0000000000000001.

Latch 0000000000000101 holds 0000000000000010, so data is copied *to* latch 0000000000000010.

Latch 0000000000000110 holds 0000000011111111, so the rightmost 8 bits of the 'to' latch are copied to.

Next, we will consider how to make a program that adds 1 to any number between 0 and 9.

First we need a way to represent the numbers 0, 1, 2, 3, 4, 5, 6, 7, 8, and 9 with only the bit values 0 and 1. The following table shows how we will do it.

Table 1

Number	representation
0	0000
1	0001
2	0010
3	0011
4	0100
5	0101
6	0110
7	0111
8	1000
9	1001

Next we need a table indicating what the answer is for each possible number 0 to 9.

Table 2

Number	answer
0	1
1	2
2	3
3	4
4	5
5	6
6	7
7	8
8	9
9	10

Next, we rewrite this table as:

Table 3

Number	answer
0	01
1	02
2	03
3	04
4	05
5	06
6	07
7	08
8	09
9	10

Next, we use table 1 to write table 3 with 1's and 0's. That is, we replace all 0's in table 3 with 0000, all 1's with 0001, and both 2's with 0010, etc. This gives us the following table.

Table 4

Number	answer
0000	00000001
0001	00000010
0010	00000011
0011	00000100
0100	00000101
0101	00000110
0110	00000111
0111	00001000
1000	00001001
1001	00010000

Next, we write table 4 as data in memory.

Table 5

label	address (number)	data (answer)	comment
add1tabl	0000000001000000	0000000000000001	
	0000000001000001	0000000000000010	
	0000000001000010	0000000000000011	
	0000000001000011	0000000000000100	
	0000000001000100	0000000000000101	
	0000000001000101	0000000000000110	
	0000000001000110	0000000000000111	
	0000000001000111	0000000000001000	
	0000000001001000	0000000000001001	
	0000000001001001	0000000000010000	

This kind of table is called a lookup table because you can look up the answer in it. *A lookup table can be made to do any function with a limited number of possible inputs.* This function is called 'increment' (add 1) and has 10 possible inputs: 0, 1, 2, 3, 4, 5, 6, 7, 8, and 9.

Next, we decide where the number to increment and the answer will be stored in memory.

label	address	data	comment
	0000000000000000	0000000000100000	start0000000000001000
	0000000000000001	0000000000001001	number to increment
	0000000000000010	0000000000000000	answer

Next, we write the program. The program is a list of instructions that tell the processor how to manipulate data. That is, the program tells the processor from where and to where copy data.

label	address	data	comment
instr_1	0000000000001000	0000000000000001	from address
	0000000000001001	0000000000001100	to address
	0000000000001010	0000000000001111	'to' bits to copy to
	0000000000001011	0000000000110000	instr.addr.and rot.
instr_2	0000000000001100	0000000001000000	from address
	0000000000001101	0000000000000010	to address
	0000000000001110	0000000011111111	'to' bits to copy to
	0000000000001111	0000000001000000	instr.addr.and rot.
instr_3	0000000000010000	0000000000000000	from address
	0000000000010001	0000000000000000	to address
	0000000000010010	0000000000000000	'to' bits to copy to
	0000000000010011	0000000001000000	instr.addr.and rot.

The whole program, including table, data, and instructions, follows.

Program to Add 1

label	address	data	comment
	0000000000000000	0000000000100000	start0000000000001000
	0000000000000001	0000000000001001	number to increment
	0000000000000010	0000000000000000	answer
instr_1	0000000000001000	0000000000000001	from address
	0000000000001001	0000000000001100	to address
	0000000000001010	0000000000001111	'to' bits to copy to
	0000000000001011	0000000000110000	instr.addr.and rot.
Instr_2	0000000000001100	0000000001000000	from address
	0000000000001101	0000000000000010	to address
	0000000000001110	0000000011111111	'to' bits to copy to
	0000000000001111	0000000001000000	instr.addr.and rot.
Instr_3	0000000000010000	0000000000000000	from address
	0000000000010001	0000000000000000	to address
	0000000000010010	0000000000000000	'to' bits to copy to
	0000000000010011	0000000001000000	instr.addr.and rot.
Add1tabl	0000000001000000	0000000000000001	
	0000000001000001	0000000000000010	
	0000000001000010	0000000000000011	
	0000000001000011	0000000000000100	
	0000000001000100	0000000000000101	
	0000000001000101	0000000000000110	
	0000000001000110	0000000000000111	
	0000000001000111	0000000000001000	
	0000000001001000	0000000000001001	
	0000000001001001	0000000000010000	

After instruction 1 (at 'instr_1') is executed, the memory has the following values. Italics show to where data was copied and arrows show from where data was copied.

After Instruction 1 Has Been Executed

label	address	data	comment
	0000000000000000	*0000000000110000*	start0000000000001000
	0000000000000001	0000000000001001	number to increment
	0000000000000010	0000000000000000	answer
instr_1	0000000000001000	0000000000000001	from address
	0000000000001001	0000000000001100	to address
	0000000000001010	0000000000001111	'to' bits to copy to
	0000000000001011	0000000000110000	instr.addr.and rot.
instr_2	0000000000001100	000000000100*1001*	from address
	0000000000001101	0000000000000010	to address
	0000000000001110	0000000011111111	'to' bits to copy to
	0000000000001111	0000000001000000	instr.addr.and rot.
instr_3	0000000000010000	0000000000000000	from address
	0000000000010001	0000000000000000	to address
	0000000000010010	0000000000000000	'to' bits to copy to
	0000000000010011	0000000001000000	instr.addr.and rot.
add1tabl	0000000001000000	0000000000000001	
	0000000001000001	0000000000000010	
	0000000001000010	0000000000000011	
	0000000001000011	0000000000000100	
	0000000001000100	0000000000000101	
	0000000001000101	0000000000000110	
	0000000001000110	0000000000000111	
	0000000001000111	0000000000001000	
	0000000001001000	0000000000001001	
	0000000001001001	0000000000010000	

After instruction 2 is executed, the memory has the following values. Italics show to where data was copied and arrows show from where data was copied.

After Instruction 2 Has Been Executed

label	address	data	comment
	0000000000000000	0000000001000000	start0000000000001000
	0000000000000001	0000000000001001	number to increment
	0000000000000010	0000000000010000	answer
instr_1	0000000000001000	0000000000000001	from address
	0000000000001001	0000000000001100	to address
	0000000000001010	0000000000001111	'to' bits to copy to
	0000000000001011	0000000000110000	instr.addr.and rot.
instr_2	0000000000001100	0000000001001001	from address
	0000000000001101	0000000000000010	to address
	0000000000001110	0000000011111111	'to' bits to copy to
	0000000000001111	0000000001000000	instr.addr.and rot.
instr_3	0000000000010000	0000000000000000	from address
	0000000000010001	0000000000000000	to address
	0000000000010010	0000000000000000	'to' bits to copy to
	0000000000010011	0000000001000000	instr.addr.and rot.
add1tabl	0000000001000000	0000000000000001	
	0000000001000001	0000000000000010	
	0000000001000010	0000000000000011	
	0000000001000011	0000000000000100	
	0000000001000100	0000000000000101	
	0000000001000101	0000000000000110	
	0000000001000110	0000000000000111	
	0000000001000111	0000000000001000	
	0000000001001000	0000000000001001	
	0000000001001001	0000000000010000	

Now, the result of 1 being added to 9 is 10 and the result, 10, written as 00010000, is stored in the last eight bits of latch 0000000000000010.

Instruction 3 ('instr_3') does nothing but execute over and over until the processor is stopped.

The program calculates what one more than nine is and finds that the answer is ten. Nine is represented as 1001 in latch 0000000000000001 and ten is represented as 00010000 in latch 0000000000000010.

Next, we will consider how to make a program that adds two numbers, from 0 to 9 together.

Again, we will represent the numbers 0 through 9 with only 1's and 0's as indicated in the following table.

Table 6

Number	Representation
0	0000
1	0001
2	0010
3	0011
4	0100
5	0101
6	0110
7	0111
8	1000
9	1001

Next, we need a table that shows the answer for each possible pair of numbers from 0 to 9. This is the addition table we studied so hard to learn in grade school and is reproduced below.

Addition Table

+	0	1	2	3	4	5	6	7	8	9
0	0	1	2	3	4	5	6	7	8	9
1	1	2	3	4	5	6	7	8	9	10
2	2	3	4	5	6	7	8	9	10	11
3	3	4	5	6	7	8	9	10	11	12
4	4	5	6	7	8	9	10	11	12	13
5	5	6	7	8	9	10	11	12	13	14
6	6	7	8	9	10	11	12	13	14	15
7	7	8	9	10	11	12	13	14	15	16
8	8	9	10	11	12	13	14	15	16	17
9	9	10	11	12	13	14	15	16	17	18

Next, we rewrite the addition table above as below.

Addition Table Listing

0 + 0 = 00
0 + 1 = 01
0 + 2 = 02
0 + 3 = 03
0 + 4 = 04
0 + 5 = 05
0 + 6 = 06
0 + 7 = 07
0 + 8 = 08
0 + 9 = 09
1 + 0 = 01
1 + 1 = 02

.
.
.

8 + 9 = 17
9 + 0 = 09
9 + 1 = 10
9 + 2 = 11
9 + 3 = 12
9 + 4 = 13
9 + 5 = 14
9 + 6 = 15
9 + 7 = 16
9 + 8 = 17
9 + 9 = 18

Only some of the table elements are listed above to save space.

Next, substituting according to table 6, we rewrite the table above as below.

Addition Table for Program

label	address	data	comment
addtable	0000001000000000	0000000000000000	0 + 0 = 00
	0000001000000001	0000000000000001	0 + 1 = 01
	0000001000000010	0000000000000010	0 + 2 = 02
	0000001000000011	0000000000000011	0 + 3 = 03
	0000001000000100	0000000000000100	0 + 4 = 04
	0000001000000101	0000000000000101	0 + 5 = 05
	0000001000000110	0000000000000110	0 + 6 = 06
	0000001000000111	0000000000000111	0 + 7 = 07
	0000001000001000	0000000000001000	0 + 8 = 08
	0000001000001001	0000000000001001	0 + 9 = 09
	0000001000010000	0000000000000001	1 + 0 = 01
	0000001000010001	0000000000000010	1 + 1 = 02
		.	
		.	
		.	
	0000001010001001	0000000000010111	8 + 9 = 17
	0000001010010000	0000000000001001	9 + 0 = 09
	0000001010010001	0000000000010000	9 + 1 = 10
	0000001010010010	0000000000010001	9 + 2 = 11
	0000001010010011	0000000000010010	9 + 3 = 12
	0000001010010100	0000000000010011	9 + 4 = 13
	0000001010010101	0000000000010100	9 + 5 = 14
	0000001010010110	0000000000010101	9 + 6 = 15
	0000001010010111	0000000000010110	9 + 7 = 16
	0000001010011000	0000000000010111	9 + 8 = 17
	0000001010011001	0000000000011000	9 + 9 = 18

Of course, in the actual program's table, *all* one hundred table elements must be included.

The table begins at 0000001000000000. It can begin anywhere in memory, just so table data doesn't overlap other data.

Next, we write the whole program, including data, instructions, and table. We want to add 9 + 7. That is, we want to calculate C=A+B where A is 9, B is 7, and C is the answer. The underlining and italics are just to highlight data for the *person* reading the program and do not affect the program. In instr_4, the first two 16-bit words of data don't matter because *no* bits are copied.

Addition Program

label	address	data	comment
start	0000000000000000	0000000000010000	start 0000000000000100
A	0000000000000001	000000000000*1001*	9 (A)
B	0000000000000010	000000000000*0111*	7 (B)
C	0000000000000011	00000000*00000000*	answer (C)
instr_1	0000000000000100	0000000000000001	from address of A
	0000000000000101	0000000000001100	to instr_3's from addr.
	0000000000000110	0000000011110000	copy to these bits
	0000000000000111	0000000000100100	go to instr_2, rot. 4
instr_2	0000000000001000	0000000000000010	from address of B
	0000000000001001	0000000000001100	to instr_3's from addr.
	0000000000001010	0000000000001111	copy to these bits
	0000000000001011	0000000000110000	go to instr_3, no rot.
instr_3	0000000000001100	00000010*00000000*	from addtable
	0000000000001101	0000000000000011	to address of C
	0000000000001110	0000000011111111	copy to these bits
	0000000000001111	0000000001000000	go to instr_4, no rot.
instr_4	0000000000010000	0000000000000000	doesn't matter
	0000000000010001	0000000000000000	doesn't matter
	0000000000010010	0000000000000000	copy NO bits
	0000000000010011	0000000001000000	go to this instruction
addtable	0000001000000000	0000000000000000	0 + 0 = 00
	0000001000000001	0000000000000001	0 + 1 = 01
	0000001000000010	0000000000000010	0 + 2 = 02
		.	
		.	
		.	
	0000001010010110	0000000000010101	9 + 6 = 15
	0000001010010111	00000000*00010110*	9 + 7 = 16
	0000001010011000	0000000000010111	9 + 8 = 17
	0000001010011001	0000000000011000	9 + 9 = 18

In the program above, instr_1 copies the value of A (9) to instr_3. Instr_2 copies the value of B (7) to instr_3. Instr_3 copies the result (16) from the 'addtable' to C. Instr_4 does nothing repeatedly. The program below shows, after the program has run, from where the data has been copied and to where the data has been copied in italics. You should try to see from where and to where the data was copied by each instruction: instr_1, instr_2, and instr_3.

After Addition Program Has Run

label	address	data	comment
start	0000000000000000	0000000000010000	start 0000000000000100
A	0000000000000001	000000000000*1001*	9 (A)
B	0000000000000010	000000000000*0111*	7 (B)
C	0000000000000011	00000000*00010110*	answer (C)
instr_1	0000000000000100	0000000000000001	from address of A
	0000000000000101	0000000000001100	to instr_3's from addr.
	0000000000000110	0000000011110000	copy to these bits
	0000000000000111	0000000000100100	go to instr_2, rot. 4
instr_2	0000000000001000	0000000000000010	from address of B
	0000000000001001	0000000000001100	to instr_3's from addr.
	0000000000001010	0000000000001111	copy to these bits
	0000000000001011	0000000000110000	go to instr_3, no rot.
instr_3	0000000000001100	00000010*10010111*	from addtable
	0000000000001101	0000000000000011	to address of C
	0000000000001110	0000000011111111	copy to these bits
	0000000000001111	0000000001000000	go to instr_4, no rot.
instr_4	0000000000010000	0000000000000000	doesn't matter
	0000000000010001	0000000000000000	doesn't matter
	0000000000010010	0000000000000000	copy NO bits
	0000000000010011	0000000001000000	go to this instruction
addtable	0000001000000000	0000000000000000	0 + 0 = 00
	0000001000000001	0000000000000001	0 + 1 = 01
	0000001000000010	0000000000000010	0 + 2 = 02
		. . .	
	0000001010010110	0000000000010101	9 + 6 = 15
	0000001010010111	00000000*00010110*	9 + 7 = 16
	0000001010011000	0000000000010111	9 + 8 = 17
	0000001010011001	0000000000011000	9 + 9 = 18

The following program multiplies two numbers together. It calculates C = A X B where A is 9, B is 7 and C is the answer (63). It is the same as the addition program except that it uses a multiplication table rather than an addition table.

Multiplication Program

label	address	data	comment
start	0000000000000000	0000000000010000	start 0000000000000100
A	0000000000000001	000000000000*1001*	9 (A)
B	0000000000000010	000000000000*0111*	7 (B)
C	0000000000000011	00000000*00000000*	answer (C)
instr_1	0000000000000100	0000000000000001	from address of A
	0000000000000101	0000000000001100	to instr_3's from addr.
	0000000000000110	0000000011110000	copy to these bits
	0000000000000111	0000000000100100	go to instr_2, rot. 4
instr_2	0000000000001000	0000000000000010	from address of B
	0000000000001001	0000000000001100	to instr_3's from addr.
	0000000000001010	0000000000001111	copy to these bits
	0000000000001011	0000000000110000	go to instr_3, no rot.
instr_3	0000000000001100	00000010*00000000*	from multiply table
	0000000000001101	0000000000000011	to address of C
	0000000000001110	0000000011111111	copy to these bits
	0000000000001111	0000000001000000	go to instr_4, no rot.
instr_4	0000000000010000	0000000000000000	doesn't matter
	0000000000010001	0000000000000000	doesn't matter
	0000000000010010	0000000000000000	copy NO bits
	0000000000010011	0000000001000000	go to this instruction
multiply	0000001000000000	0000000000000000	0 X 0 = 00
	0000001000000001	0000000000000000	0 X 1 = 00
	0000001000000010	0000000000000000	0 X 2 = 00
		.	
		.	
		.	
	0000001010010110	0000000001010100	9 X 6 = 54
	0000001010010111	00000000*01100011*	9 X 7 = 63
	0000001010011000	0000000001110010	9 X 8 = 72
	0000001010011001	0000000010000001	9 X 9 = 81

Notice that one can save a lot of work by salvaging (copying) instructions from a program one has already written for use in a new, similar program. Tables can often be salvaged as well.

The following program adds two two-digit numbers, A (99) and B (87), together for a result of C (186). First, it adds the right digits together (9+7) for a result of 16. 16 is 6 with a carry. Then, the carry, 1, is added to 9 and 8 for a result of 18. That makes the entire answer 186. Adding 1+9+8 together requires an add *with carry,* so we need a table with carry of 1 or 0 as below. For this table, there are 200 possibilities. There are 2 values of carry (0 or 1), 10 values of one input (0-9), and 10 values of another input (0-9) for 2 X 10 X 10 = 200 possibilities. Notice the carry, '+0' and '+1,' in the upper left of the tables below. The two tables below are two halves of the entire table.

Add Two Digits Program

label	address	data	comment
start	0000000000000000	0000000000010000	start 0000000000000100
A	0000000000000001	0000000010011001	99 (A)
B	0000000000000010	0000000010000111	87 (B)
C	0000000000000011	0000000000000000	answer (C)
instr_1	0000000000000100	0000000000000001	from address of A
	0000000000000101	0000000000001100	to instr_3's from addr.
	0000000000000110	0000000011110000	copy to these bits
	0000000000000111	0000000000100100	go to instr_2, rot. 4
instr_2	0000000000001000	0000000000000010	from address of B
	0000000000001001	0000000000001100	to instr_3's from addr.
	0000000000001010	0000000000001111	copy to these bits
	0000000000001011	0000000000110000	go to instr_3, no rot.
instr_3	0000000000001100	0000001000000000	from addtable
	0000000000001101	0000000000000011	to address of C
	0000000000001110	0000000011111111	copy to these bits
	0000000000001111	0000000001000000	go to instr_4, no rot.
instr_4	0000000000010000	0000000000000001	from address of A
	0000000000010001	0000000000011100	to instr_7's from addr.
	0000000000010010	0000000011110000	copy to these bits
	0000000000010011	0000000001010000	go to instr_5, no rot.
instr_5	0000000000010100	0000000000000010	from address of B
	0000000000010101	0000000000011100	to instr_7's from addr.
	0000000000010110	0000000000001111	copy to these bits
	0000000000010111	0000000001101100	to instr_6,rot.4 right
instr_6	0000000000011000	0000000000000011	from addr.of C (carry)
	0000000000011001	0000000000011100	to instr_7's from addr.
	0000000000011010	0000000100000000	copy to this bit
	0000000000011011	0000000001110100	to instr_7,rot.4 left
instr_7	0000000000011100	0000001000000000	from addtable
	0000000000011101	0000000000000011	to address of C
	0000000000011110	0000111111110000	copy to these bits
	0000000000011111	0000000010000100	go to instr_8, rot. 4
instr_8	0000000000100000	0000000000000000	doesn't matter
	0000000000100001	0000000000000000	doesn't matter
	0000000000100010	0000000000000000	copy NO bits
	0000000000100011	0000000010000000	go to this instruction
addtable	0000001000000000	0000000000000000	0 + 0 + 0 = 00
	0000001000000001	0000000000000001	0 + 0 + 1 = 01
	0000001000000010	0000000000000010	0 + 0 + 2 = 02
		. . .	
	0000001010010110	0000000000010101	0 + 9 + 6 = 15
	0000001010010111	0000000000010110	0 + 9 + 7 = 16
	0000001010011000	0000000000010111	0 + 9 + 8 = 17
	0000001010011001	0000000000011000	0 + 9 + 9 = 18
	0000001100000000	0000000000000001	1 + 0 + 0 = 01
	0000001100000001	0000000000000010	1 + 0 + 1 = 02
	0000001100000010	0000000000000011	1 + 0 + 2 = 03

.
.
.

0000001110010110	0000000000010110	1 + 9 + 6 = 16
0000001110010111	0000000000010111	1 + 9 + 7 = 17
0000001110011000	00000000*00011000*	1 + 9 + 8 = 18
0000001110011001	0000000000011001	1 + 9 + 9 = 19

In some high level languages, instructions 1 through 7 can be written with one instruction, 'C = A + B.' You type in 'C = A + B.' Then you run another program that is called a compiler. The compiler converts 'C = A + B' into all that machine language, instr_1 through instr_7. A compiler can greatly ease writing programs. Writing programs in machine language (1's and 0's) is relatively difficult. (Most other processors have a hardware adder, so C = A + B becomes few instructions though many bits are added.)

The program below shows the result of running the program above. To and from where data has been copied is underlined *and* in italics. The answer is 000110000110, or 186, and is stored in the word (16 bits) labeled 'C.' The first word (the first 16 data bits) of instr_3 now holds 1001,0111 or 9,7 and the first word of instr_7 now holds 1,1001,1000 or 1,9,8, where the 1 in 1,9,8 is represented by only 1 bit (1).

After Add Two Digits Program Has Run

label	address	data	comment
start	0000000000000000	0000000010000000	start 0000000000000100
A	0000000000000001	00000000*10011001*	99 (A)
B	0000000000000010	00000000*10000111*	87 (B)
C	0000000000000011	0000*000110000110*	answer (C)
instr_1	0000000000000100	0000000000000001	from address of A
	0000000000000101	0000000000001100	to instr_3's from addr.
	0000000000000110	0000000011110000	copy to these bits
	0000000000000111	0000000000100100	go to instr_2, rot. 4
instr_2	0000000000001000	0000000000000010	from address of B
	0000000000001001	0000000000001100	to instr_3's from addr.
	0000000000001010	0000000000001111	copy to these bits
	0000000000001011	0000000000110000	go to instr_3, no rot.
instr_3	0000000000001100	00000010*10010111*	from addtable
	0000000000001101	0000000000000011	to address of C
	0000000000001110	0000000011111111	copy to these bits
	0000000000001111	0000000001000000	go to instr_4, no rot.
instr_4	0000000000010000	0000000000000001	from address of A
	0000000000010001	0000000000011100	to instr_7's from addr.
	0000000000010010	0000000011110000	copy to these bits
	0000000000010011	0000000001010000	go to instr_5, no rot.
instr_5	0000000000010100	0000000000000010	from address of B
	0000000000010101	0000000000011100	to instr_7's from addr.
	0000000000010110	0000000000001111	copy to these bits
	0000000000010111	0000000001101100	to instr_6,rot.4 right
instr_6	0000000000011000	0000000000000011	from addr.of C (carry)
	0000000000011001	0000000000011100	to instr_7's from addr.
	0000000000011010	0000000100000000	copy to this bit
	0000000000011011	0000000001110100	to instr_7,rot.4 left
instr_7	0000000000011100	0000001*110011000*	from addtable

	0000000000011101	0000000000000011	to address of C
	0000000000011110	0000111111110000	copy to these bits
	0000000000011111	0000000010000100	go to instr_8, rot. 4
instr_8	0000000000100000	0000000000000000	doesn't matter
	0000000000100001	0000000000000000	doesn't matter
	0000000000100010	0000000000000000	copy NO bits
	0000000000100011	0000000010000000	go to this instruction
addtable	0000001000000000	0000000000000000	0 + 0 + 0 = 00
	0000001000000001	0000000000000001	0 + 0 + 1 = 01
	0000001000000010	0000000000000010	0 + 0 + 2 = 02
	.		
	.		
	.		
	0000001010010110	0000000000010101	0 + 9 + 6 = 15
	0000001010010111	000000000*0010110*	0 + 9 + 7 = 16
	0000001010011000	0000000000010111	0 + 9 + 8 = 17
	0000001010011001	0000000000011000	0 + 9 + 9 = 18
	0000001100000000	0000000000000001	1 + 0 + 0 = 01
	0000001100000001	0000000000000010	1 + 0 + 1 = 02
	0000001100000010	0000000000000011	1 + 0 + 2 = 03
		.	
		.	
		.	
	0000001110010110	0000000000010110	1 + 9 + 6 = 16
	0000001110010111	0000000000010111	1 + 9 + 7 = 17
	0000001110011000	000000000*0011000*	1 + 9 + 8 = 18
	0000001110011001	0000000000011001	1 + 9 + 9 = 19

The program below includes an example of 'branching.' Branching in a program means that either some instructions or some other instructions are executed depending on a value (usually a bit) in memory (or, in most processors, in a register). Branch bits are often called flags. In the program below, the rightmost bit in latch A (at address 0000000000000001) is a flag and determines whether instr_4 or instr_5 is executed. Instr_4 copies the 'all 1's pattern' (1111111111111111) in latch C to latch B. Instr_5 copies the '1,0 pattern' (1010101010101010) from latch D to latch B. If latch A has value 1, then instr_5 is executed. If A has value 0, then instr_4 is executed. In the program below, A has value 1 so instr_5 is executed and B gets the '1,0 pattern.'

Instr_3 does nothing but go to the next instruction because it copies *no* bits. Instr_3 has 00000000010*0*0000 in latch (address) 0000000000001111, so, normally, instr_4, at address 0000000000010000, would be executed next. However, instr_2 copies the rightmost bit (1) from A into the fifth-from-rightmost bit of latch 0000000000001111 so that latch 0000000000001111 contains 00000000010*1*0000 and instr_5, at address 0000000000010100, is executed after instr_3. The instructions are then executed in the following order: instr_2, instr_3, instr_5, instr_6. Notice that instr_4 is skipped.

Branching Program

label	address	data	comment
start	0000000000000000	0000000000100000	start 0000000000001000
A	0000000000000001	000000000000000*1*	flag(1 or 0)right bit
B	0000000000000010	*0000000000000000*	value to change
C	0000000000000011	1111111111111111	all 1's pattern
D	0000000000000100	*1010101010101010*	1,0 pattern
instr_2	0000000000001000	0000000000000001	from A
	0000000000001001	0000000000001111	to next of instr_3
	0000000000001010	0000000000010000	change this bit
	0000000000001011	0000000000110100	to instr_3, rot.4
instr_3	0000000000001100	0000000000000000	doesn't matter
	0000000000001101	0000000000000000	doesn't matter
	0000000000001110	0000000000000000	copy NO bits
	0000000000001111	0000000001*0*00000	to instr_4 OR instr_5
instr_4	0000000000010000	0000000000000011	from C pattern
	0000000000010001	0000000000000010	to B
	0000000000010010	1111111111111111	copy all bits
	0000000000010011	0000000001100000	to instr_6, no rot.
instr_5	0000000000010100	0000000000000100	from D pattern
	0000000000010101	0000000000000010	to B
	0000000000010110	1111111111111111	copy all bits
	0000000000010111	0000000001100000	to instr_6, no rot.
instr_6	0000000000011000	0000000000000000	doesn't matter
	0000000000011001	0000000000000000	doesn't matter
	0000000000011010	0000000000000000	copy NO bits
	0000000000011011	0000000001100000	go to this instruction

The result of running the program above is shown below. Notice that B now holds 1010101010101010 and that address 0000000000001111 now holds 00000000010*1*0000 instead of 00000000010*0*00000. Italics show to where and from where values have been copied.

After Branching Program Has Run

label	address	data	comment
start	0000000000000000	*0000000001100000*	start 0000000000001000
A	0000000000000001	000000000000000*1*	flag(1 or 0)right bit
B	0000000000000010	*1010101010101010*	value to change
C	0000000000000011	1111111111111111	all 1's pattern
D	0000000000000100	*1010101010101010*	1,0 pattern
instr_2	0000000000001000	0000000000000001	from A
	0000000000001001	0000000000001111	to next of instr_3
	0000000000001010	0000000000010000	change this bit
	0000000000001011	0000000000110100	to instr_3, rot.4
instr_3	0000000000001100	0000000000000000	doesn't matter
	0000000000001101	0000000000000000	doesn't matter
	0000000000001110	0000000000000000	copy NO bits

	0000000000001111	0000000001000000	to instr_4 OR instr_5
instr_4	0000000000010000	0000000000000011	from C pattern
	0000000000010001	0000000000000010	to B
	0000000000010010	1111111111111111	copy all bits
	0000000000010011	0000000001100000	to instr_6, no rot.
instr_5	0000000000010100	0000000000000100	from D pattern
	0000000000010101	0000000000000010	to B
	0000000000010110	1111111111111111	copy all bits
	0000000000010111	0000000001100000	to instr_6, no rot.
instr_6	0000000000011000	0000000000000000	doesn't matter
	0000000000011001	0000000000000000	doesn't matter
	0000000000011010	0000000000000000	copy NO bits
	0000000000011011	0000000001100000	go to this instruction

The following program is the same as the previous one (before it was run) except that A now has value 0 instead of value 1 (in the rightmost bit). This means that instr_4 will be executed instead of instr_5 and B will get 1111111111111111 from C rather than 1010101010101010 from D. The instructions are executed in the following order: instr_2, instr_3, instr_4, instr_6. Instr_5 is not executed.

Branching Program with Flag = 0

label	address	data	comment
start	0000000000000000	0000000000100000	start 0000000000001000
A	0000000000000001	000000000000000*0*	flag(1 or 0)right bit
B	0000000000000010	*0000000000000000*	value to change
C	0000000000000011	1111111111111111	all 1's pattern
D	0000000000000100	1010101010101010	1,0 pattern
instr_2	0000000000001000	0000000000000001	from A
	0000000000001001	0000000000001111	to next of instr_3
	0000000000001010	0000000000010000	change this bit
	0000000000001011	0000000000110100	to instr_3, rot.4
instr_3	0000000000001100	0000000000000000	doesn't matter
	0000000000001101	0000000000000000	doesn't matter
	0000000000001110	0000000000000000	copy NO bits
	0000000000001111	0000000001000000	to instr_4 OR instr_5
instr_4	0000000000010000	0000000000000011	from C pattern
	0000000000010001	0000000000000010	to B
	0000000000010010	1111111111111111	copy all bits
	0000000000010011	0000000001100000	to instr_6, no rot.
instr_5	0000000000010100	0000000000000100	from D pattern
	0000000000010101	0000000000000010	to B
	0000000000010110	1111111111111111	copy all bits
	0000000000010111	0000000001100000	to instr_6, no rot.
instr_6	0000000000011000	0000000000000000	doesn't matter
	0000000000011001	0000000000000000	doesn't matter
	0000000000011010	0000000000000000	copy NO bits
	0000000000011011	0000000001100000	go to this instruction

The following program shows the result of executing the preceding program. Notice that B now contains 1111111111111111 from C and address 0000000000001111 still contains 0000000001000000. Instr_4 has been executed instead of instr_5.

After Branching Program with Flag = 0 Has Run

label	address	data	comment
start	0000000000000000	0000000001100000	start 0000000000001000
A	0000000000000001	0000000000000000	flag(1 or 0)right bit
B	0000000000000010	1111111111111111	value to change
C	0000000000000011	1111111111111111	all 1's pattern
D	0000000000000100	1010101010101010	1,0 pattern
instr_2	0000000000001000	0000000000000001	from A
	0000000000001001	0000000000001111	to next of instr_3
	0000000000001010	0000000000010000	change this bit
	0000000000001011	0000000000110100	to instr_3, rot.4
instr_3	0000000000001100	0000000000000000	doesn't matter
	0000000000001101	0000000000000000	doesn't matter
	0000000000001110	0000000000000000	copy NO bits
	0000000000001111	0000000001000000	to instr_4 OR instr_5
instr_4	0000000000010000	0000000000000011	from C pattern
	0000000000010001	0000000000000010	to B
	0000000000010010	1111111111111111	copy all bits
	0000000000010011	0000000001100000	to instr_6, no rot.
instr_5	0000000000010100	0000000000000100	from D pattern
	0000000000010101	0000000000000010	to B
	0000000000010110	1111111111111111	copy all bits
	0000000000010111	0000000001100000	to instr_6, no rot.
instr_6	0000000000011000	0000000000000000	doesn't matter
	0000000000011001	0000000000000000	doesn't matter
	0000000000011010	0000000000000000	copy NO bits
	0000000000011011	0000000001100000	go to this instruction

MISCELLANEOUS

Computer with Input and Output

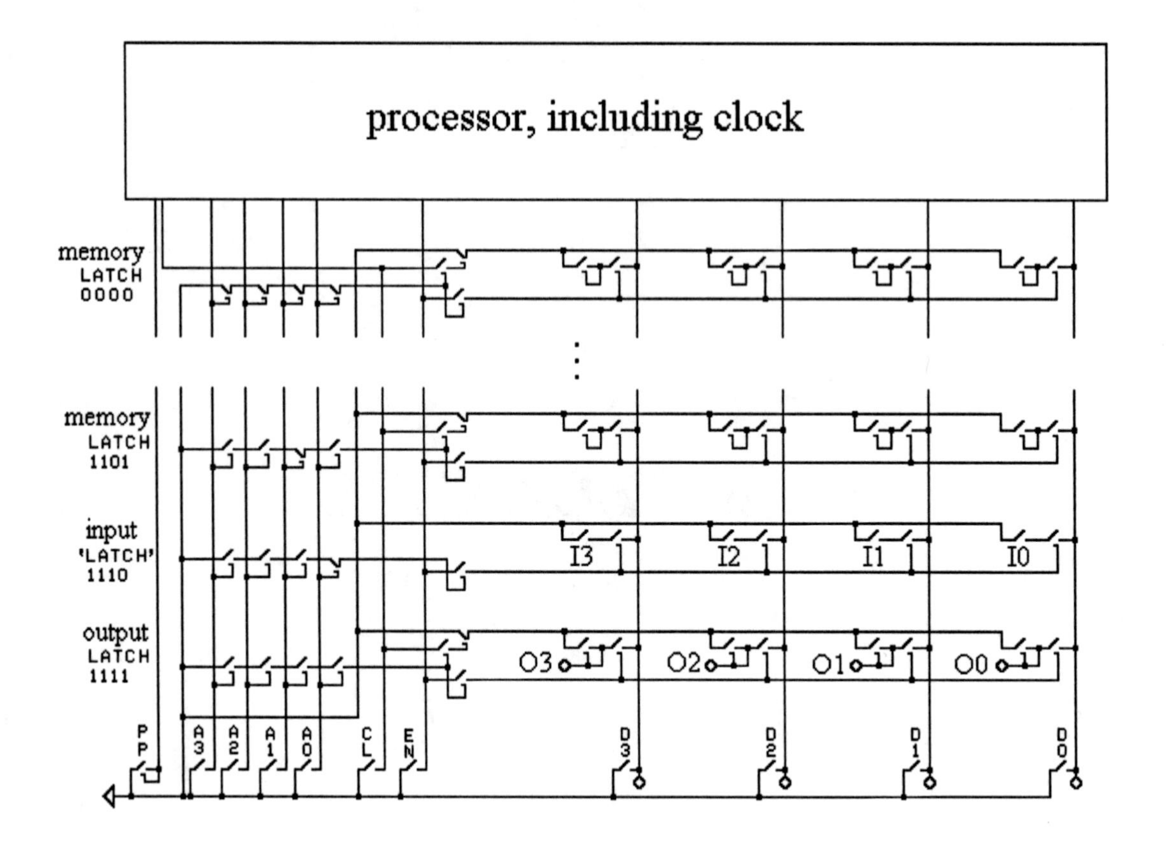

Inputs and outputs have been added to the computer above in place of two memory latches. When data is written to (copied to) 'output latch 1111,' then each loop, O3, O2, O1, and O0, will turn on its light if a 1 is stored in the loop. When data is read from (copied from) input 'latch' 1110 (It's not really a latch because it doesn't have loops.), then a 1 will be copied from key I3 if key I3 is pressed. If key 3 is *not* pressed, then a 0 is copied from I3. I2, I1, and I0 work similarly.

For example, the following one-instruction program (for the four-bit computer above) copies data from address (input 'latch') 1110 to address (output latch) 1111 over and over again. Therefore, when the program is running, pressing key I3 turns on light O3, pressing key I2 turns on light O2, pressing key I1 turns on light O1, and pressing key I0 turns on light O0.

label	address	data	comment
start	0000	0100	start at 0100
instr_1	0100	1110	from inputs
.	0101	1111	to outputs
.	0110	1111	copy all bits
.	0111	0100	repeat this instruction, no rotate

The keys PP, A3, A2,…D0 along the bottom of the computer diagrammed above allow you to control the computer. You can write to memory, start the processor, stop the processor, and read the results from memory. These keys are, together, called the control panel. A control panel controlled early computers. However, today a keyboard controls a computer. A keyboard is a lot of keys similar to the input keys. The computer runs a program that checks for key presses and reacts accordingly. That program is called an operating system. A joystick may control a game computer. Inside a typical joystick are keys that the joystick bumps into. Those keys and the keys under the joystick's buttons are also like the input keys above. The outputs can control motors (like in a disk drive) rather than lights.

Transistors

Modern computers use two types of transistors, which correspond to the two types of relays. An N-channel transistor corresponds to a normally open relay. A P-channel transistor corresponds to a normally closed relay. However, transistors have some idiosyncrasies and you can't simply replace relays with transistors to make a successful transistor-based design. It takes about twice as many transistors as relays to do something. Of course, the high speed and low cost of transistors make transistors vastly superior in spite of the extra design effort required. The millions of transistors in a modern microprocessor allow for more than one type of instruction. For example, besides, or instead of, rotate and mask, the instruction set can include add, subtract, multiply, divide, etc.

The Future

This completes the explanation of how the vast majority of computers work now. One instruction is executed at a time. In modern designs, it is common for the ensuing instruction to be started before the prior instruction finishes, so a few instructions can be executed at once. There are designs that allow many instructions to be executed at once; but such computers, though very fast, are relatively hard to program and, mainly for that reason, have not become very popular. Most such designs use many (often relatively simple) computers, each of which can execute an instruction at once, and which communicate with each other through inputs and outputs. Computers with such designs are called parallel computers and are probably what will be used in the future. For example, I have an idea for a computer that would be able to execute thousands of instructions at a time and still be programmed almost the same way as a normal computer. That will have to be the subject of another book.

About the Author

Roger Stephen Young lives in Pennsylvania and graduated from The Pennsylvania State University where he majored in physics and was interested in transistors. He went to the California State University at Fullerton and worked on a Master's degree in electrical engineering for two years, but got a job at Texas Instruments before finishing. He has extensive programming experience and is currently promoting his parallel processor design that can be programmed easily and has a novel inter-processor communication architecture.

Printed in the United States
1516900001B/208

9 781403 32582